Recent Research and Advances in Soilless Culture

Edited by Metin Turan, Sanem Argin, Ertan Yildirim and Adem Güneş

Published in London, United Kingdom

Recent Research and Advances in Soilless Culture
http://dx.doi.org/10.5772/intechopen.96844
Edited by Metin Turan, Sanem Argin, Ertan Yildirim and Adem Güneş

Contributors
Moaed Ali Al Meselmani, Duraisamy Kalaivanan, Govindan Selvakumar, Arockiasamy Carolin Rathinakumari, Torkwase Emmanuella Bulya, Tatiana V. Glukhareva, Elena G. Kovaleva, Genaro Martin Soto-Zarazúa, Arantza Elena Sánchez-Gutiérrez, Juan Fernando García-Trejo, Manuel Toledano-Ayala, Ristina Siti Sundari, Lies Sulistyowati, Trisna Insan Noor, Iwan Setiawan, Víctor García-Gaytán, Fanny Hernández-Mendoza

First published in London, United Kingdom, 2023 by IntechOpen
IntechOpen is the global imprint of INTECHOPEN LIMITED, registered in England and Wales, registration number: 11086078, 5 Princes Gate Court, London, SW7 2QJ, United Kingdom

British Library Cataloguing-in-Publication Data
A catalogue record for this book is available from the British Library

Additional hard and PDF copies can be obtained from orders@intechopen.com

Recent Research and Advances in Soilless Culture
Edited by Metin Turan, Sanem Argin, Ertan Yildirim and Adem Güneş
p. cm.
Print ISBN 978-1-80355-168-5
Online ISBN 978-1-80355-169-2
eBook (PDF) ISBN 978-1-80355-170-8

Meet the editors

Metin Turan received his Ph.D. from the Department of Soil Science and Plant Nutrition, Atatürk University, Turkey, in 2002. He is currently a professor in the Department of Genetics and Bioengineering, at Yeditepe University, Turkey. His research focuses on soil ecology and biological fertilizer applications. He has more than 100 research publications to his credit. Prof. Turan is a scientific member of many international organizations such as IFOAM, FIBL, and EBIC, and he has chaired many conferences in Turkey and Europe.

Ertan Yildirim is a full professor in the Department of Horticulture, at Atatürk University, Turkey. He obtained a Ph.D. in Horticulture from the same university in 2003. His research focuses on vegetable growing, vegetable breeding, greenhouse management, seed germination and physiology, organic agriculture, and stress physiology. He has more than 200 publications to his credit.

Sanem Argin received her Ph.D. in Food Science from the University of Maryland, College Park, USA, in 2007. She is currently head of the Department of Agricultural Trade and Management, at Yeditepe University, Turkey. Her research focuses on biostimulant applications for sustainable agriculture and biopolymer applications in food science and agriculture. She has published many journal articles, book chapters, and one patent.

Adem Güneş is an associate professor in the Department of Soil Science and Plant Nutrition, at Erciyes University, Turkey. He received his Ph.D. in Soil Science and Plant Nutrition from Atatürk University Erzurum, Turkey, in 2013. He is focused on teaching and researching plant nutrition and fertilization, soil chemistry, plant stress conditions, plant biochemistry, soil sanity, fertilization, humus chemistry, plant breeding, and phytoremediation. He is the author of more than 100 papers published in international journals. Dr. Güneş has served as a reviewer for many journals and has attended many international congresses and symposiums.

Contents

Preface

The Food and Agriculture Organization (FAO) of the United Nations (UN) has reported that thirty football fields of soil are lost annually. It takes 1000 years to produce only 3 cm of soil. At present, the soil is the most common media used for growing crops throughout the world, as it provides the support, nutrition, air, and water required for the optimum growth of plants. However, the soil has some serious limitations, mainly in safe food production, due to climate change (temperature, solar intensity, water deficiency, soil pollution, low biodiversity, soil degradation, soil salinity), soil-borne diseases, nematodes, drainage, transport, short life cycle, and other factors. The challenges facing soil-based agriculture have led to the invention of soilless agriculture. A plant does not need soil to grow, but it does need soil to obtain nutrients. Soilless cultivation has become possible through the use of advanced technology. Soilless farming refers to growing a wide range of crops, particularly horticultural crops, in different growing media or substrates other than soil (substrate culture) or in an aerated nutrient solution (water culture). In other words, it includes all systems of plant growth either on porous substrates or on the pure nutrient solution (NS) instead of natural soil. Soilless culture practices worldwide emerged with the idea that existing agricultural land will not meet the food needs of the rapidly increasing population. Soilless culture practices, which have significant advantages over traditional methods of agriculture, are increasing globally.

Soilless culture can be defined as an advanced production technique in which plant cultivation is carried out using different solid or liquid media other than soil, in the root zone of the plant nutrients and water required for the development of plants. The soilless culture model aims to provide the most appropriate balance of air, water, and nutrients in the root zone in an artificial environment, as well as physical support for plants.

There are many reasons for the emergence of soilless culture. The most important ones are the increasing population and land-related problems. It is estimated that the world population, which is 6.5 billion today, will reach 9 billion in 2050 and will need 60–70 percent more food. In many countries, the lands that can be opened to agriculture have reached their last limit. It is increasingly common to move soils out of agriculture for reasons such as erosion, aridity, and settlement. Soilless culture has emerged as an important alternative, especially in countries with insufficient soil and water resources, to eliminate foreign dependency in terms of food, to meet the food needs of the increasing population in general, and to obtain greater yields in terms of water and fertilizers. The prohibition of methyl bromide, which has been widely used in the fight against soil-borne diseases and nematodes in recent years, has been effective in the spread of soilless culture.

Soilless agriculture can be performed in controlled environments to address many of the concerns we now have. That is why it is expected to be the future method of farming in many parts of the world. It helps in the intensive production of crops under full or partially controlled conditions. It guarantees flexibility and intensification of crop production systems in areas with adverse growing conditions. It also helps with precise control over the supply of water and nutrients. Soilless farming furthermore helps eliminate soil-borne diseases, reduces labor requirements, and increases annual crop yield. Hence, it is becoming very popular in urban and semi-urban areas. In fact, it has become the method of choice for places where there are abundant and frequent soil-borne diseases. At present, it has been widely used to produce different vegetables.

This book provides an update on the sources, production, and applications of soilless agriculture and highlights their importance in terms of sustainable agriculture, biodiversity, and the environment. It presents reliable, complete, and up-to-date information on soil history, industry, and production. It also addresses soilless culture types in the world, plant nutrition and fertilizer use activities, plant diseases and protection techniques, carbon and water footprint, organic cultivation in soilless agriculture, connection of management with smart agricultural tools, and evaluation of economic aspects of soilless culture.

Metin Turan and Sanem Argin
Yeditepe University,
Istanbul, Turkey

Ertan Yildirim
Atatürk University,
Erzurum, Turkey

Adem Güneş
Erciyes University,
Kayseri, Turkey

Chapter 1

The Effects of Different Substrates with Chemical and Organic Fertilizer Applications on Vitamins, Mineral, and Amino Acid Content of Grape Berries from Soilless Culture

Serpil Tangolar, Semih Tangolar, Metin Turan, Mikail Atalan and Melike Ada

Abstract

Due to its advantages, soilless cultivation has been used for both early- and late-maturing grape varieties. High nutritional and energy value is one of the strongest features that make the grape an effective component of agriculture and the human diet. Therefore, it was thought that it would be useful to determine the nutrient content of the berries in a soilless culture study carried out on the Early Cardinal grape variety. One-year-old vines were trained to a guyot system and grown in 32-liter plastic pots containing four different solid growing media, namely, zeolite, cocopeat, and zeolite+cocopeat (Z + C) (1:1 and 1:2, v:v). A total of three different nutrient solutions (Hoagland, Hoagland A (adapted to the vine) and organic liquid worm fertilizer (OLWF)) were applied to the plants. Grapevines were given different solutions starting from the bud burst. Z + C (1:1) substrate mixture giving the highest values of 14 amino acids, vitamins, and most macro- and microelements. Hoagland and Modified Hoagland nutrient solutions mostly gave higher values than OLWF for the properties studied. In general, it was observed that there were no significant losses in terms of mineral, vitamin, and amino acid composition in soilless grape cultivation.

Keywords: grapevine, phytochemicals, fertilization, vermicompost, zeolite, cocopeat

1. Introduction

Grapes (*Vitis vinifera* L.) are the most produced fruit in the world. The total grape area and its production globally are 7.4 million ha and 77.8 million tons, respectively, in 2018 [1]. About 36% of the total is consumed for fresh, 7% for dried, and 57% for winemaking. Five countries represent 50% of the world's vineyards. Turkey is in the fifth position in vineyard areas in the world in 2018 with a total surface of 448,000 ha, after Spain, China, France, and Italy. It is the sixth in total grape production (3.9 million tons) among the major grape producers that after China, Italy, USA, Spain, and France; fourth in table grapes (2.2 million tons, 56.1%), and first in dried grape production (396,825 tons, 40.7%), about fortieth in wine grape production among the grape-growing countries. In Turkey, the grapes used for winemaking are 124,800 tons (3.2%) [1].

Soilless culture techniques are primarily applied in ornamental plants and vegetables in the world and Turkey [2, 3]. In recent years, this technique is also used to overcome some problems due to its various advantages in grape cultivation [2, 4–6]. No need for tillage and soil preparation, protection from soil pathogens, effective use of water and nutrient solutions, reduction of spraying, obtaining more quantity and quality products per unit area, production of new or traditional grape varieties in a more extended period according to market demands, and control of harvest time are among some advantages of soilless cultivation [2, 4, 7].

In the world and Turkey, when it is considered together with the cultivation of greenhouse grapes for early grape ripening or late harvest, grape cultivation in soilless culture is considered an important cultivation method due to its advantages. This technique may be used for both early- and late-maturing grape varieties. According to our current information, no producer grows grapes commercially in soilless culture in Turkey. Studies on the subject are still carried out in horticulture departments of some agriculture faculties and viticulture research institutes.

Depending on the research purposes, different varieties, substrate mixtures, containers and nutrient solutions [2, 4, 7–15] were used in the grape cultivation experiments in the soilless culture system.

In the studies conducted by Tangolar et al. [6], the effect of substrates on the grape yield and quality of the berries in vines grown in the open and under the greenhouse was determined. The study that examined the yield, cluster, and berry properties of Early Sweet variety determined that perlite:peat (2:1) and cocopeat substrates gave better results. Tangolar et al. [16] also researched Early Sweet and Trakya Ilkeren cultivars to determine the effects of three different media, namely perlite:peat (2:1), cocopeat and pumice, and two different modified Hoagland nutrient solutions on shoot diameter as well as the nutrient element and chlorophyll levels of the leaves and grape yield and quality characteristics. The study found a significant difference between media and nutrient solution application for some characteristics examined.

Achieving a good quality in grapes is an essential goal wherever it is grown; one of the important components that make up the quality is the phytochemical content of the berries. Grapes contain a number of phytochemicals beneficial for human health, as well as amino acids, proteins, vitamins, and minerals [17–26]. So, berries are efficiently used to increase the nutritional and energy value of the human diet.

Some studies [27] have shown that magnesium, calcium, zinc, and vitamins such as B and C are related to people's cognitive performance. Clinical findings have revealed that extreme deficiencies of one or more of these nutrients are not uncommon, even in developed countries. These deficiencies may affect cognitive performance, especially in vulnerable groups such as the elderly and those exposed to occupational pressures and difficult living conditions.

Key et al. [28] noted that dietary science is increasingly recognized for its ability to prevent and support disease prevention and new technologies and therapies to improve modern medical practice. Researchers noted that dietary studies help discover specific dietary patterns that promote healthy brain aging and moderate the involvement of nervous systems known to facilitate cognitive performance in later life [28].

The composition of grape berries in different grape cultivars grown open field is affected by different factors such as variety, stress conditions, biostimulants, irrigation, fertigation, pruning, and others [26, 29–49].

In spite of this, the studies conducted in the world and Turkey found no study of the effects of the different substrates and nutrition solutions on the biochemical content of berries obtained from varieties grown in soilless culture. So, this subject is thought to have not been sufficiently investigated yet.

Because of these, it has been seen beneficial to examine the effects of substrates and nutrition solutions on the biochemical contents, which are essential for human health. Therefore, this study was designated to evaluate the amino acid, mineral, and vitamin content of berries from Early Cardinal table grape cultivar grown in different soilless culture medium and plant nutrient solutions.

2. Materials and methods

2.1 Trial conditions

This research was carried out in a greenhouse at the Department of Horticulture, Faculty of Agriculture, the University of Cukurova, which was conducted under a 21 m, 9 m, and 3 m in length, width, and height greenhouse covered with UV plastic with a thickness of 0.4 mm. During the research, no heating process was done in the greenhouse.

2.2 Plant material

As plant material, own-rooted Early Cardinal grape (*V. vinifera* L.) cv. grown in soilless culture was used. To produce plant material, cuttings from Early Cardinal grapes (*V. vinifera* L.) grown were planted in perlite pools on January 15, 2018, and irrigated immediately after planting. Rooting of cutting occurred after approximately 90 days at a satisfactory level. Well-rooted cuttings were selected and transplanted into 32-liter plastic pots containing four different solid growing media, namely, zeolite, cocopeat, zeolite+cocopeat (Z + C) (1:1, v:v), and Z + C (1:2, v:v). A total of three different nutrient solutions were applied to the rooted cuttings: two chemical nutrient solutions (Hoagland (H) and Hoagland A (HA- adapted to the vine) and organic liquid worm fertilizer (OLWF) (**Table 1**). Nitrogen, phosphorus, potassium, magnesium, sulfur, and boron concentrations in the modified Hoagland solution were reduced between 3.2% (phosphorus) and 76.5% (sulfur) compared with Hoagland, and on the other hand, iron 2, manganese 6, zinc 20, and molybdenum 5 fold have been increased. With the same amount of solution in volume, more N, P, Mg, Zn, Cu, Mn, and Fe were given than Hoagland A and Hoagland through OLWF. The pots were placed in the greenhouse with a distance of 1.50 m between rows and 0.60 m in rows. After planting, a well-irrigation was performed to saturate the cultivation media.

One-year-old vines entered the resting period at the end of the first year were pruned and trained to a guyot system to prepare for the crop year, on January 31, 2019. About 20 buds were left per vine. The number of clusters of the vines was equal to 12 clusters by removing the excessive clusters on May 24, 2019, after the berry set. Grapevines were given different solutions within the second vegetation year, starting from the bud burst.

The pH value of the tap water used in the experiment was 7.68, and the EC value was 0.813 mS cm^{-1}. The amount of water given to the plants varied between 1 and

Element	Formula	Hoagland A (mg kg^{-1})	Hoagland (mg kg^{-1})	Organic liquid worm fertilizer
N	$K_2(NO_3)_2$	150	210	5%
P	H_3PO_4	30	31	0.49%
K	K_2SO_4	175	235	1.47%
Mg	$MgSO_4.7H_2O$	20	48	0.78%
S	$CaSO_4.H_2O$	15	64	Not detected
Fe	Fe-EDDHA	5	2.5	5257 ppm
Mn	$MnSO_4.\ H_2O$	3	0.5	565 ppm
B	H_3BO_3	0.4	0.5	Not detected
Cu	$CuSO_4\ 5H_2O$	0.02	0.02	58 ppm
Zn	$ZnSO_4\ 7H_2O$	1	0.05	152.5 ppm
Mo	$(NH_4)_6Mo_7O_{24}.4\ H_2O$	0.05	0.01	Not detected
pH				5.28
Total dry matter				13%
Humic-fulvic acid				38%

Table 1.
Composition and formula of chemical and organic nutrient solutions used in the trial.

3 L pot^{-1} per day according to the water-holding capacity of the growth medium. The total amount of nutrients applied per plant in the first and crop year of the experiment is shown in **Table 2**.

2.3 Biochemical analysis

When the total soluble solids (TSS) reached about 12–14%, five cluster samples were taken from each of the three replicates of treatments on July 1, 2019. After removing from the clusters, stored berries at −20°C before the phytochemical analysis were analyzed in the Department of Genetic and Bio-Engineering, Faculty of Engineering, University of Yeditepe.

2.3.1 Mineral elements

Macro and micronutrient element analyses were carried out using samples of berries. Phosphorus (P) was determined vanadomolibdo phosphoric acid yellow color method as reported by Bremner [50]. Potassium (K), calcium (Ca), magnesium (Mg), iron (Fe), zinc (Zn), copper (Cu), and manganese (Mn) concentrations of the berries were analyzed by atomic absorption spectrophotometer [51].

2.3.2 Amino acids

1 g fresh sample was treated with 0.1 N HCl, homogenized with ultra turrax, and incubated at 4°C for 12 hours. Supernatants were filtered through 0.22-m filters after

Element	Hoagland A		Hoagland		Organic liquid worm fertilizer	
	2018	2019	2018	2019	2018	2019
N (g)	12.75	21.00	17.85	29.39	37.40	59.90
P (g)	2.55	4.20	2.64	4.34	3.67	5.87
K (g)	14.87	24.50	19.97	32.89	10.99	17.61
Mg (g)	1.89	15.91	4.53	37.47	5.83	9.34
Zn (mg)	84.92	139.86	4.165	6.86	114.07	182.70
Cu (mg)	1.70	2.80	1.70	2.80	43.38	69.48
B (mg)	85.0	140.00	106.25	175.00	Not detected	Not detected
Mn (mg)	255.0	420.00	42.5	70.00	422.62	676.87
Mo (mg)	0.43	0.70	0.09	0.14	Not detected	Not detected
Fe (mg)	474.8	777.9	235.5	387.8	3932.2	6297.9

Table 2.
The amount of nutrients given per plant by different nutrient solutions in 2 years.

samples were centrifuged at 1200 rpm for 50 minutes (Millex Millipore). The supernatants were then transferred to a vial, and the amino acids were analyzed in HPLC as described by Antoine et al. [52] and Kitir et al. [53]. Readings from Zorbax Eclipse-AAA 4.6150 mm and 3.5 m columns (Agilent 1200 HPLC) were taken at 254 nm, and the amino acids were identified by comparing them to standards of O-phthaldialdehyde (OPA), fluorenylmethyl-chloroformate (FMOC), and 0.4 N borate. The following solutions were used in the mobile phase chromatography system: Phase A: 40 mM NaH_2PO_4 (pH: 7.8) and Phase B: acetonitrile/methanol/water (45/45/10 v/v/v), after a 26-minute derivation process in HPLC, aspartate, glutamate, asparagine, serine, glutamine, histidine, glycine, arginine, alanine, tyrosine, cysteine, valine, methionine, tryptophan, phenylalanine, isoleucine, leucine, lysine, thionine, and proline.

A 50 mg frozen berry sample was crushed using liquid nitrogen and extracted with 4.5 mL of 3-sulfosalicylic acid, and then filtered through a Whatman filter paper (#2) for proline measurement. In a test tube, 2 mL of the filtrate were mixed with 2 mL acid-ninhydrin and 2 mL glacial acetic acid for 1 hour at 100°C, stopped the reaction with an ice bath, and the filtrates were analyzed. The concentration of proline was measured spectrophotometrically at 520 nm [54].

2.3.3 Vitamins

2.3.3.1 Vitamin A

Berry samples were ground for vitamin A (Retinol). Berry samples were extracted with a mixture of n-hexane and ethanol. 1% BHT was added and kept in the dark environment for 1 day. At the end of this period, centrifugation was conducted at 4000 rpm (+4°C) for 10 min. The obtained supernatant was filtered with the help of Whatman filter paper and added 0.5 mL of n-hexane. Drying was then performed using nitrogen gas. The residue in the tubes was dissolved in a methanol + tetrahydrofuran mixture. Analyses were carried out in Thermo Scientific Finnigan Surveyor model high-performance liquid chromatography (HPLC) and in amber glass vials on Tray, and autosampler using PDA array detector [55, 56].

2.3.3.2 Vitamin B

A total of 10 g of samples were weighed and homogenized. The samples were then transferred to a conical flask with 25 mL of extraction solution. A shaking water bath at an ambient temperature of 70°C was used to sonicate the solution for 40 minutes. Following sonication, the sample was cooled and filtered to make a volume of 50 mL with extraction solution. The extraction solution was again filtered with filter trips (0.45 μm), and 20 μl aliquots solution was injected into the HPLC by using an auto-sampler. A reversed-phase C-18 analytical column (STR ODS-M, 150 mm 4.6 mm ID, 5 m, Shimadzu Corporation, Japan) separated the B complex vitamins. At 40°C, the mobile phase consists of a 9:1 (v/v) combination of 100 mM sodium phosphate buffer (pH: 2.2) containing 0.8 mM sodium-1-octane sulfonate and acetonitrile. The flow rate was constant at 0.8 mL/min using a PDA detector with a 270 nm absorption rate. The peak area of the corresponding chromatogram was used to calculate B vitamins using the following equation [57]:

$$\begin{aligned} \text{B vitamins}\left(\text{mg}100\,\text{g}^{-1}\right) = {} & \text{Concentration of standard} \\ & \text{x}\left(\text{Area of sample / Area of Standard}\right) \\ & \text{x Dilution factor} \end{aligned}$$

2.3.3.3 Vitamin C

Plants were sliced, frozen in liquid nitrogen, and kept at a temperature of −80°C until the analyses were completed. The extraction solution was combined with 2.5 ml of frozen crushed plant material (3% MPA and 8% acetic acid for MPA-acetic acid extraction and 0.1% oxalic acid for oxalic acid extraction). The mixture was titrated with indophenol solution (25% DCIP and 21% $NaHCO_3$ in water) until light, but the distinct rose-pink color appeared and persisted for more than 5 seconds [58].

2.4 Experimental design and statistical analysis

The study was designed according to the "Randomized Complete Blocks" with three replicates in 12 treatments. For each application and replicate, approximately 500 g of the berry samples were taken and analyzed for the compounds to be studied. Data obtained from the study were subjected to variance analysis using the SAS-based JMP statistical package programmer. The least significant difference (LSD) test was used to separate different groups at a 5% significance level.

3. Results and discussions

Besides bodywork, vitamins, and minerals, protection of the body from diseases, blood formation, bone, dental health, etc., are required for functions. Each food contains different amounts of various vitamins and minerals. Its richest sources are fresh vegetables and fruits [59].

As shown in **Table 3**, there were significant differences among the substrates related to macro- and microelements of berries except for boron. Considering, P, K, Ca, Mg, Mn, and Cu concentrations of berries were higher in Z + C (1:1) than the other substrates. However, zeolite, cocopeat, and Z + C (1:1) for Na, Cocopeat, and Z + C (1:1) for Fe, and zeolite for Zn concentrations gave higher values than the other applications. Phosphorus, Mg, Fe in Hoagland; K in Hoagland A; calcium, Na, and Mn in Hoagland and Hoagland A, and zinc in OLWF fertilizers were recorded have higher concentrations than those of the others.

Macrominerals presented in **Table 3** determined that the potassium contents of berries were higher than those of the others, ranging from 234 mg 100 g^{-1} for Z + C

Sources of variation	Macroelements (mg 100 g^{-1})				
	P	K	Ca	Mg	Na
Substrate					
Zeolite	17.7 c[y]	213 b	48 b	13.7 d	2.7 a
Cocopeat	19.1 b	208 c	47 b	17.9 b	2.4 a
Z + C (1:1)[x]	21.0 a	234 a	51 a	20.0 a	2.4 a
Z + C (1:2)	15.4 d	193 d	39 c	16.7 c	1.9 b
LSD 5%	0.4	5	2	0.8	0.3
p-value	<0.0001	<0.0001	<0.0001	<0.0001	0.0011
Fertilizer					
Hoagland A	19.3 b	227 a	49 a	16.8 b	2.6 a
Hoagland	19.8 a	223 b	50 a	18.1 a	2.6 a
OLWF	15.8 c	186 c	40 b	16.3 b	1.9 b
LSD 5%	0.4	4	1	0.7	0.3
p-value	<0.0001	<0.90001	<0.0001	<0.0001	<0.0001
Interaction					
Zeolite × Hoagland A	2.52 a	3.35a	0.67 a	1.61de	0.43 a
Zeolite × Hoagland	1.63 ef	1.92 f	0.46 d	1.25 g	0.29 b
Zeolite × OLWF	1.15 ı	1.13 j	0.31 g	1.24 g	0.08 e
Cocopeat × Hoagland A	1.38 h	1.45 ı	0.36 f	1.41 f	0.16 d
Cocopeat × Hoagland	2.31 c	2.24 d	0.55 b	1.97b	0.28 b
Cocopeat × OLWF	2.06 d	2.54 c	0.50 c	1.98b	0.27 b
Z + C (1:1) × Hoagland A	2.40 b	2.48 c	0.57 b	2.20a	0.26 bc
Z + C (1:1) × Hoagland	2.34 bc	2.87 b	0.56 b	2.26a	0.27 b
Z + C (1:1) × OLWF	1.55 g	1.67 h	0.39 e	1.56e	0.20 cd
Z + C (1:2) × Hoagland A	1.40 h	1.81 g	0.36 f	1.49ef	0.19 d
Z + C (1:2) × Hoagland	1.65 e	1.88 fg	0.41 e	1.77c	0.19 d
Z + C (1:2) × OLWF	1.57 fg	2.10 e	0.40 e	1.74 cd	0.19 d
LSD 5%	0.7	8	3	1.3	0.6
p-value	<0.0001	<0.0001	<0.0001	<0.0001	<0.0001

[x]*Z + C: Zeolite+Cocopeat, OLWF: Organic liquid worm fertilizer,*
[y]*Mean separation within columns by LSD multiple range test at 0.05 level.*

Table 3.
The effect of different substrates and nutrient solution applications on the level of macro elements in berries.

(1:1) substrate and 186 mg 100 g^{-1} for OLWF fertilizer. Followed calcium content of grapes was found between 51 mg 100 g^{-1} for Z + C (1:1) substrate and 40 mg 100 g^{-1} for OLWF fertilizer. Among the macroelements, sodium gave the lowest amount.

Considering trace elements, the highest iron content (0.362 mg 100 g^{-1}) is obtained from Z + C (1:1), whereas the lowest level of iron (0.255 mg 100 g^{-1}) was found in zeolite. The zinc content of grape berries was in the range of 0.299 and 0.184 mg 100 g^{-1}, whereas the manganese content of grape berries was in the range of 0.235–0.178 mg 100 g^{-1}. Cupper and boron microminerals varied between 0.147 and 0.105 and 0.481 and 0.329 mg 100 g^{-1}, respectively. The substrate × fertilizer interaction was significant for all elements except Cu and B (**Tables 3** and **4**).

In the study by Abdrabba and Hussein [35], calcium, magnesium, potassium, phosphorus, and iron values were determined as 120, 31, 154, 39, and 5 mg 100 g^{-1} as the average of pulp, seed, and peel, respectively, and these minerals useful for the human body have been deemed necessary.

Similarly, the values given in Kral et al. [59] for Ca, K, Mg, Na, Cu, Fe, Mn, and Zn; in Cantürk et al. [60] for Ca, K, Mg, Na, P, Cu, Fe, Mn, B, and Zn; in Abdrabba and Hussein [35] for Ca, K, Mg, P, and Fe; in Anonymous [61] for Ca, K, Mg, Na, and Fe; in Olsen and Ware [62] for Ca, K, Mg, Na, P, Fe, Mn, B, and Zn were found to be quite close to the values given in **Table 3** for the specified elements.

For this reason, it was concluded that there were no significant losses in terms of mineral levels of grapes grown under soilless culture conditions.

Vitamins, like minerals, are micronutrients that play an essential role in fulfilling metabolic functions, producing new cells, and repairing damaged cells.

There were found significant differences among substrates and fertilizers in terms of vitamin contents of berries analyzed in the study. The higher vitamin A, B1, B2, B6, and C values were analyzed in berries of plants grown in Z + C (1:1) substrate mix and berries of applications using Hoagland solution (**Table 5**). The higher values obtained from vitamin A, B1, B2, B6, and C were 39.21, 65.12, 167.06, 95.19, and 15.21 mg 100 g^{-1}, respectively. The substrate × fertilizer interaction was significant for all vitamins examined (**Table 5**).

According to the Bourre [63] and Key et al. [28], nutrients such as vitamins, minerals, and amino acids play a crucial role in ensuring proper brain function. Vitamins protect against inflammation and reactive oxidative species. Minerals function as cofactors for enzymes, prevent lipid peroxidation, and promote energy production. Amino acids serve as precursors to neurotransmitters and neuromodulator metabolites responsible for various functions related to attention, mood, arousal, and memory.

Most vitamins and microelements have been studied concerning brain functioning. For example, it has been reported by Bourre [63] that the use of glucose for energy production occurs in the presence of vitamin B1. This vitamin regulates cognitive performance, especially in the elderly. It has been reported that vitamin B6 is beneficial in treating premenstrual depression. Vitamins B6 and B12, among others, are directly involved in synthesizing certain neurotransmitters. Vitamin B12 delays the onset of signs of dementia and blood abnormalities when administered at an appropriate time before the first symptoms.

Emphasizing the importance of mineral nutrients for healthy brain aging, Key et al. [28] stated in their results that a nutrient regime containing macro- and micronutrients softens the effect of brain structure on cognitive function in old age and supports the effectiveness of interdisciplinary methods in nutritional cognitive neuroscience for a healthy brain. In the article of Çetin et al. [64], different researchers reported

Sources of variation	Microelements (mg 100 g^{-1})				
	Fe	Zn	Mn	Cu	B
Substrate					
Zeolite	0.255 c [y]	0.299 a	0.178 c [y]	0.105 b	0.348
Cocopeat	0.353 a	0.184 c	0.208 b	0.131ab	0.448
Z + C (1:1)[x]	0.362 a	0.187 c	0.235 a	0.147 a	0.481
Z + C (1:2)	0.288 b	0.192 b	0.195 b	0.113 ab	0.329
LSD 5%	0.011	0.011	0.016	0.036	NS
p value	<0.0001	<0.0001	<0.0001	0.1082	0.002
Fertilizer					
Hoagland A	0.325 b	0.206 b	0.208 a	0.123	0.399
Hoagland	0.340 a	0.207 b	0.216 a	0.136	0.455
OLWF	0.279 c	0.233 a	0.188 b	0.112	0.351
LSD 5%	0.010	0.009	0.014	NS	NS
p-value	<0.0001	<0.0001	0.001	0.2907	0.3459
Interaction					
Zeolite × Hoagland A	373.26 c	23.36c	257.02 b	111.36	33.55
Zeolite × Hoagland	274.67e	26.09b	161.89 fg	107.69	36.95
Zeolite × OLWF	119.72 g	40.33a	115.50 h	96.29	33.83
Cocopeat × Hoagland A	229.96f	22.09 cd	145.29 g	113.61	38.77
Cocopeat × Hoagland	399.01 b	17.68gh	222.25 cd	159.97	59.94
Cocopeat × OLWF	430.45 a	15.31ı	255.55 b	120.14	35.54
Z + C (1:1) × Hoagland A	403.44 b	19.74ef	247.77 bc	177.22	61.79
Z + C (1:1) × Hoagland	404.49 b	17.81gh	290.87 a	135.89	40.02
Z + C (1:1) × OLWF	276.79de	18.58fgh	166.47 fg	126.40	42.59
Z + C (1:2) × Hoagland A	294.99 d	17.26 h	182.54 f	90.53	25.58
Z + C (1:2) × Hoagland	282.14de	21.29de	188.47 ef	142.21	44.89
Z + C (1:2) × OLWF	289.78de	19.16 fg	212.86 de	104.91	28.33
LSD 5%	0.020	0.018	0.028	NS	NS
p-value	<0.0001	<0.0001	<0.0001	0.3888	0.3886

[x]*Z + C: Zeolite+Cocopeat, OLWF: Organic liquid worm fertilizer.*
[y]*Mean separation within columns by LSD multiple range test at 0.05 level,*
NS: Nonsignificant.

Table 4.
The effect of different substrates and nutrient solution applications on the level of microelements in berries.

that potassium is a very important component of human health. A high-potassium diet lowers blood pressure and reduces cardiovascular disease morbidity and mortality [65]. In addition, potassium intake reduces urinary calcium excretion and decreases the risk of osteoporosis [66]. Ca is the primary element of the bone system, assists in tooth development, helps regulate endo- and exo-enzymes, and plays a significant role in regulating blood pressure [67]. Therefore, it is an essential mineral for human

Sources of variation	A Retinol	B1 Thiamin	B2 Riboflavin	B6 Pyridoxine	C Ascorbic acid
Substrate					
Zeolite	29.95 d [y]	45.39 b	113.76 d	78.50 c	12.49 c
Cocopeat	34.91 b	59.59 a	148.49 b	88.27 b	13.51 b
Z + C (1:1)[x]	39.21 a	65.12 a	167.06 a	95.18 a	15.21 a
Z + C (1:2)	31.65 c	46.02 b	121.29 c	69.74 d	12.14 c
LSD 5%	1.09	5.54	6.59	4.55	0.42
p-value	<0.0001	<0.0001	<0.0001	<0.0001	<0.0001
Fertilizer					
Hoagland A	34.51 b	55.67 b	140.93 b	84.44 b	13.62 b
Hoagland	36.51 a	60.47 a	153.29 a	91.79 a	14.46 a
OLWF	30.76 c	45.95 c	118.74 c	72.54 c	11.93 c
LSD 5%	0.95	4.80	5.71	3.94	0.36
p-value	<0.0001	<0.0001	<0.0001	<0.0001	<0.0001
Interaction					
Zeolite × Hoagland A	39.40 b	56.80 bc	144.69 de	93.26 b	15.72 b
Zeolite × Hoagland	28.89 de	49.01 cd	114.02 fg	80.73 c	12.41 d
Zeolite × OLWF	21.54 g	30.37e	82.57 h	61.52 f	9.33 f
Cocopeat × Hoagland A	26.70 f	43.21 d	106.56 g	71.01 de	10.88 e
Cocopeat × Hoagland	39.49 b	74.24 a	187.54 b	109.98 a	15.58 b
Cocopeat × OLWF	38.53 b	61.32 b	151.37 d	83.81 c	14.07 c
Z + C (1:1) × Hoagland A	43.75 a	82.81 a	204.58 a	113.18 a	17.43 a
Z + C (1:1) × Hoagland	43.59 a	63.66 b	172.08 c	94.21 b	16.08 b
Z + C (1:1) × OLWF	30.29 d	48.88 cd	124.53 f	78.14 cd	12.11 d
Z + C (1:2) × Hoagland A	28.19 ef	39.86 de	107.89 g	60.31 f	10.43 e
Z + C (1:2) × Hoagland	34.08 c	54.98 bc	139.50 e	82.23 c	13.76 c
Z + C (1:2) × OLWF	32.67 c	43.22 d	116.47 fg	66.69 ef	12.21 d
LSD 5%	1.89	9.60	11.41	7.88	0.72
p-value	<0.0001	<0.0001	<0.0001	<0.0001	<0.0001

[x]Z + C: Zeolite+Cocopeat, OLWF: organic liquid worm fertilizer.
[y]Mean separation within columns by LSD multiple range test at 0.05 level.

Table 5.
The effect of different substrate and nutrient solution applications on vitamins (mg 100 g^{-1}).

health. Zn and Fe deficiency in the diet programs is a common problem and a matter of great concern, especially in developing countries where people trust vegetarian diets more. Zn is involved with the immune system, and Fe is concerned with hemoglobin, myoglobin, and cytochrome [68]. They are also recognized to be potential antioxidants [69]. Mg is essential to all living cells, where they play a major role in manipulating important biological polyphosphate compounds such as ATP, DNA, and RNA. Also, more than 300 enzymes require magnesium ions to function [70].

In the study, the effects of applications on 20 amino acids in grapes were evaluated. For all amino acids examined in **Table 5**, the differences between treatments were statistically significant. The highest values were found from Z + C (1:1) application in 14 amino acids (**Table 6**), namely aspartate, glutamate, proline, arginine, glutamine, histidine, alanine, cystine, methionine, tryptophan, phenylalanine, isoleucine, leucine, and lysine. In Z + C (1:1), Z + C (1:2), and cocopeat applications for valine; in Z + C (1:1) and zeolite for serine; and in cocopeat and Z + C (1:2) applications for glycine were the highest values. Apart from these, the highest tyrosine and asparagine in Zeolite were detected. Among nutrient solutions, Hoagland for aspartate, glutamate, alanine, and phenylalanine amino acids; Hoagland and Hoagland A for proline, arginine, glutamine, tyrosine, methionine, tryptophan, isoleucine, and leucine; Hoagland and OLWF nutrient solutions for histidine; Hoagland A for glycine, thionine, cystine, valine, lysine, asparagine and serine amino

Sources of Variation	Aspartate	Glutamate	Proline	Arginine	Glutamine
Substrate					
Zeolite	14,930 c [y]	10,637 d	28,607 c	34,258 c	20,750 c
Cocopeat	16,289 b	14,849 b	33,667 b	39,258 b	24,768 b
Z + C (1:1)[x]	17,718 a	15,751 a	37,901 a	42,880 a	27,569 a
Z + C (1:2)	13,867 d	12,257 c	34,200 b	35,427 c	22,018 c
LSD 5%	5529	774	1290	2222	1668
p-value	<0.0001	<0.0001	<0.0001	<0.0001	<0.0001
Fertilizer					
Hoagland A	16,172 b	13,440 b	34,041 a	39,771 a	24,293 a
Hoagland	16,725 a	15,096 a	34,020 a	38,911 a	25,437 a
OLWF	14,206 c	11,585 c	32,720 b	35,186 b	21,599 b
LSD 5%	470	670	1117	1924	1445
p-value	<0.0001	<0.0001	0.0342	0.0001	<0.0001
Interaction					
Zeolite × Hoagland A	20,134 ab	14,265 c	42,259 c	51,443 a	26,212 bc
Zeolite × Hoagland	13,650 efg	12,818 de	22,751 ıj	28,563 ef	19,198 ef
Zeolite × OLWF	11,005 ı	4828 g	20,810 j	22,769 g	16,841 f
Cocopeat × Hoagland A	12,168 h	10,323 f	23,521 ı	26,383 fg	18,822 ef
Cocopeat × Hoagland	18,646 cd	18,030 a	32,766 f	40,354 c	28,919 ab
Cocopeat × OLWF	18,052 d	16,195 b	44,713 b	51,038 a	26,562 b
Z + C (1:1) × Hoagland A	19,396 bc	17,604 a	36,692 e	46,293 b	31,632 a
Z + C (1:1) × Hoagland	20,511 a	16,144 b	51,120 a	54,359 a	30,277 a
Z + C (1:1) × OLWF	13,248 fg	13,505 cd	25,890 h	27,989 f	20,799 de
Z + C (1:2) × Hoagland A	12,990 gh	11,568 ef	33,693 f	34,966 d	20,506 de
Z + C (1:2) × Hoagland	14,091 ef	13,390 cd	29,442 g	32,367 de	23,354 cd
Z + C (1:2) × OLWF	14,520 e	11,814 e	39,465 d	38,948 c	22,193 d
LSD 5%	940	1341	2234	3849	2889
p-value	<0.0001	<0.0001	<0.0001	<0.0001	<0.0001

Sources of variation	Histidine	Glycine	Thionine	Alanine	Tyrosine
Substrate					
Zeolite	1895 d	2190 b	5423 a	22,905 c	2724 a
Cocopeat	3454 b	2510 a	5598 a	26,921 b	2535 bc
Z + C (1:1)[x]	3752 a	2200 b	4870 b	30,365 a	2632 ab
Z + C (1:2)	3113 c	2560 a	5699 a	25,722 b	2455 c
LSD 5%	243	150	289	1855	138
p-value	<0.0001	<0.0001	<0.0001	<0.0001	0.0034
Fertilizer					
Hoagland A	2892 b	2710 a	6197 a	26,486 ab	2807 a
Hoagland	3149 a	2130 c	4904 b	27,826 a	2689 a
OLWF	3119 a	2260 b	5091 b	25,123 b	2264 b
LSD 5%	211	130	250	1607	120
p-value	0.073	<0.0001	<0.0001	0.0079	<0.0001
Interaction					
Zeolite × Hoagland A	2314 fg	141.2 e	4365 ef	29,162 cd	4232 a
Zeolite × Hoagland	1313 h	169.9 d	4589 e	20,585 fg	2817 c
Zeolite × OLWF	2059 g	346.6 ab	7314 bc	18,968 g	1124 g
Cocopeat × Hoagland A	2360 fg	367.8 a	7761 ab	20,839 fg	1900 f
Cocopeat × Hoagland	3648 c	157.1 de	3686 gh	28,825 cd	2623 cd
Cocopeat × OLWF	4355 b	227.4 c	5348 d	31,100 bc	3082 b
Z + C (1:1) × Hoagland A	3761 c	337.4 b	7120 c	32,508 b	2561 d
Z + C (1:1) × Hoagland	4904 a	150.8 de	3484 h	35,810 a	3072 b
Z + C (1:1) × OLWF	2592 f	170.7 d	4005 fg	22,776 f	2263 e
Z + C (1:2) × Hoagland A	3134 de	235.6 c	5541 d	23,435 ef	2535 d
Z + C (1:2) × Hoagland	2732 ef	372.2 a	7856 a	26,085 de	2243 e
Z + C (1:2) × OLWF	3472 cd	160.0 de	3699 gh	27,646 d	2589 cd
LSD 5%	422	260	501	3214	239
p-value	<0.0001	<0.0001	<0.0001	<0.0001	<0.0001
Sources of variation	**Cysteine**	**Valine**	**Methionine**	**Tryptophan**	**Phenylalanine**
Substrate					
Zeolite	3846 ab[y]	1526 b	6339 c	5409 c	7410 d
Cocopeat	3675 b	1728 a	7544 b	5845 b	9456 b
Z + C (1:1)[x]	3995 a	1892 a	8232 a	6663 a	10,707 a
Z + C (1:2)	3272 c	1805 a	6697 c	5886 b	8196 c
LSD 5%	177	170	599	329	595
p-value	<0.0001	0.0015	<0.0001	<0.0001	<0.0001
Fertilizer					
Hoagland A	3986 a	1818 a	7405 a	6213 a	9070 b
Hoagland	3822 b	1655 b	7501 a	6018 a	9796 a
OLWF	3283 c	1740 ab	6702 b	5621 b	7961 c
LSD 5%	153	147	519	285	515

Sources of variation	Cysteine	Valine	Methionine	Tryptophan	Phenylalanine
p-value	<0.0001	0.0930	0.0079	0.0010	<0.0001
Interaction					
Zeolite × Hoagland A	6100 a	2834 b	9659 b	8966 a	9836 de
Zeolite × Hoagland	3273 f	934 e	5259 f	4190 g	7250 gh
Zeolite × OLWF	2164 j	810 e	4099 g	3071 h	5146 i
Cocopeat × Hoagland A	2525 i	920 e	4934 fg	3689 g	6936 h
Cocopeat × Hoagland	3975 d	1410 d	8014 c	5578 e	11,157 bc
Cocopeat × OLWF	4523 c	2854 b	9685 b	8268 b	10,276 cd
Z + C (1:1) × Hoagland A	4098 d	1454 d	8261 c	6291 d	12,360 a
Z + C (1:1) × Hoagland	5093 b	3214 a	10,906 a	9520 a	11,623 ab
Z + C (1:1) × OLWF	2794 hi	1007 e	5528 f	4178 g	8139 fg
Z + C (1:2) × Hoagland A	3220 fg	2065 c	6766 de	5906 de	7150 gh
Z + C (1:2) × Hoagland	2945 gh	1062 e	5827 ef	4785 f	9157 ef
Z + C (1:2) × OLWF	3650 e	2288 c	7496 cd	6967 c	8285 f
LSD 5%	307	294	1038	571	1031
p-value	<0.0001	<0.0001	<0.0001	<0.0001	<0.0001
Sources of Variation	**Isoleucine**	**Leucine**	**Lysine**	**Asparagine**	**Serine**
Substrate					
Zeolite	4933 c	9161 c	7862 c	9618 a	16,332 a
Cocopeat	5582 ab	10,046 b	9003 b	7140 c	14,232 b
Z + C (1:1)[x]	6111 a	11,322 a	9860 a	8111 b	15,996 a
Z + C (1:2)	5119 bc	9917 bc	9350 ab	8500 b	14,284 b
LSD 5%	531	790	658	754	1060
p-value	0.0006	0.0001	<0.0001	<0.0001	0.0003
Fertilizer					
Hoagland A	5717 a	10,580 a	9411 a	9851 a	16,941 a
Hoagland	5528 a	10,270 a	8620 b	7332 b	15,112 b
OLWF	5064 b	9485 b	9024 ab	7844 b	13,580 c
LSD 5%	460	684	570	653	918
p-value	0.0214	0.0092	0.0297	<0.0001	<0.0001
Interaction					
Zeolite × Hoagland A	7633 a	14,380 ab	14,573 b	20,483 a	28,776 a
Zeolite × Hoagland	3996 ef	7216 de	4845 fg	5060 fg	12,623 fg
Zeolite × OLWF	3170 f	5889 e	4168 g	3310 h	7599 j
Cocopeat × Hoagland A	3672 ef	6456 e	4777 fg	3636 h	9376 ij
Cocopeat × Hoagland	5610 bc	10,072 c	7385 e	5030 fg	13,807 ef
Cocopeat × OLWF	7463 a	13,609 b	14,846 b	12,755 c	19,514 c
Z + C (1:1) × Hoagland A	6440 b	11,145 c	7614 e	5672 f	15,296 de
Z + C (1:1) × Hoagland	7999 a	15,692 a	16,718 a	14,686 b	22,072 b

Sources of Variation	Isoleucine	Leucine	Lysine	Asparagine	Serine
Z + C (1:1) × OLWF	3894 ef	7129 de	5247 fg	3974 gh	10,621 hı
Z + C (1:2) × Hoagland A	5122 cd	10,338 c	10,683 d	9611 e	14,315 ef
Z + C (1:2) × Hoagland	4507 de	8099 d	5531 f	4552 fgh	11,949 gh
Z + C (1:2) × OLWF	5728 bc	11,316 c	11,835 c	11,338 d	16,588 d
LSD 5%	919	1369	1140	1305	1836
p-value	<0.0001	<0.0001	<0.0001	<0.0001	<0.0001

[x]Z + C: Zeolite+Cocopeat, OLWF: organic liquid worm fertilizer.
[y]Mean separation within columns by LSD multiple range test at 0.05 level.

Table 6.
The effect of different substrate and nutrient solution applications on amino acid content (μg kg^{-1}) of Early Cardinal berries.

acids gave the highest values. As can be seen in **Table 6**, substrate × fertilizer interaction was found to be significant for all amino acids.

Proline is reported in many works of literature as an amino acid whose synthesis is increased, especially under abiotic stress conditions such as drought [43, 71]. For this reason, we evaluated that the high increase in proline amino acid in Hoagland A and Hoagland nutrient solutions may be due to the lower amounts of some macro- (N) and microelements (Zn, Cu, Mn, Fe) in these solutions compared with OLWF nutrient solution (**Table 1**). Anjum et al. [72], Liang et al. [73], and Arabshahi and Mobasser [74] indicated that sensitive plants are less able to accumulate solutes, but increases in proline can be found in most organisms (including animals) following water stress [25, 43].

According to the Huang and Ough [29], Canoura et al. [43], Bouzas-Cid et al. [36, 47–49], Sánchez-Gómez et al. [41], Gutiérrez-Gamboa et al. [26, 42, 45, 46], Fernández-Novales et al. [75], and Wu et al. [44], amino acid contents of grape berries are affected by different variety, rootstock, location and fertilization, etc., viticultural practices. For instance, in the study by Gutiérrez-Gamboa et al. [26], the effect of foliar application of a seaweed extract to a Tempranillo Blanco variety on must and wine amino acids and ammonium content was determined. The results suggested that Tempranillo Blanco behaved as an arginine accumulator variety. Biostimulation after seaweed applications at a high dosage to the grapevines increased the concentration of several amino acids in the 2017 season while scarcely affecting their content in 2018.

In the another research by Gutiérrez-Gamboa et al. [46], results showed that of some elicitors and nitrogen foliar applications to Garnacha and Tempranillo grapevines decreased the must amino acid concentration. The treatments applied to Graciano grapevines affected the grape amino acid content. According to the percentage of variance attributable, the variety had a higher effect on the must amino acid composition than the treatments and their interaction. In the study by Fernández-Novales et al. [75], researchers have investigated the use of visible and near-infrared spectroscopy to estimate the grape amino acid content on whole berries of Grenache grape variety. Amino acid values ranged between 0.01 mg L^{-1} (Leucine) and 341 mg L^{-1} (Arginine). In their results, amino acid values obtained in our study varied from 1526 μg kg^{-1} (valine in zeolite) to 42,880 μg kg^{-1} (arginine in Z + C (1:1)).

These values were close to the values of valine (1.07 mg L^{-1}) given by Fernández-Novales et al. [75] for Grenache and arginine (38.44–89.60 mg L^{-1}) given by Valdes et al. [76] for Tempranillo berries. Arginine and proline amino acids were recorded as the most abundant amino acids in all media and nutrient solutions used in our experiment; valine, glycine, and tyrosine were determined as the amino acids with the lowest values. These results agree with Fernández Novales et al. [75] and Valdes et al. [76] that arginine and proline were also reported as the most abundant amino acids, both of the researches.

From the above statements, it has been concluded that grapes grown in soilless culture will not encounter a significant nutrient loss in terms of amino acids examined in this study. In our study, it has been evaluated that the Z + C (1:1) mixture substrate, which has the higher values for 14 amino acids, including proline as well as arginine, is remarkable in terms of nutrient saving.

4. Conclusions

According to the main results obtained from this study;

- In soilless culture cultivation of table grapes, it has been observed that zeolite and cocopeat media can be used alone, as well as a 1:1 mixture of Zeolite:Cocopeat, where the highest values are obtained.

- Hoagland and modified Hoagland nutrient solutions mostly gave higher values than OLWF for the properties studied. However, since OLWF did not have a significant negative effect, it was considered that it would be appropriate to continue working with this and similar solutions.

- Amino acid, vitamins, and mineral contents of grapes grown in soilless culture conditions were found to be close to the values given in the literature for grapes grown in open field.

Acknowledgements

This article was produced from the Master Thesis of Mikail Atalan, whose study was supported by the Cukurova University Scientific Research Coordination Unit (Project No: FYL-2018-11066).

Author details

Serpil Tangolar[1]*, Semih Tangolar[1], Metin Turan[2], Mikail Atalan[1] and Melike Ada[1]

1 Faculty of Agriculture, Department of Horticulture, University of Cukurova, Adana, Türkiye

2 Faculty of Engineering, Department of Genetic and Bio-Engineering, University of Yeditepe, Istanbul, Türkiye

*Address all correspondence to: stangolar@cu.edu.tr

References

[1] OIV. 2019 Statistical Report on World Vitiviniculture [Internet]. 2019. Available from: OIV-2019-statistical-report-on-world-vitiviniculture.pdf [Accessed: November 19, 2021]

[2] Di Lorenzo R, Pisciotta A, Santamaria P, Scariot V. From soil to soil-less in horticulture: Quality and typicity. Italian Journal of Agronomy. 2013;**8**(4):255-260

[3] Gül A. Topraksız tarım (3. Baskı). İzmir: Hasad Yayıncılık; 2019. p. 146

[4] Buttaro D, Serio F, Santamaria P. Soilless greenhouse production of table grape under Mediterranean conditions. Journal of Food, Agriculture & Environment. 2012;10(2): 641-645. Available from: https://www.researchgate.net/profile/Pietro-Santamaria/publication/256547182_Soilless_greenhouse_production_of_table_grape_under_Mediterranean_conditions/links/569368bd08aee91f69a7ee14/Soilless-greenhouse-production-of-table-grape-under-Mediterranean-conditions.pdf

[5] Tangolar S, Kaya S, Alkan Torun A, Tarım G, Ada M, Aydın O. The effects of different growing media and root pruning applications on Trakya Ilkeren grape variety grown in soilless culture. Yüzüncü Yil Üniversitesi Journal of Agricultural Sciences. 2018;**28**(Special Issue):321-328

[6] Tangolar S, Baştaş PC, Alkan Torun A, Tangolar S. Effects of substrate and crop load on yield and mineral nutrition of Early Sweet grape cultivar grown in soilless culture. Erwerbs-obstbau. 2019;**61**:33-40. DOI: 10.1007/s10341-019-00446-0

[7] Di Lorenzo R, Dimauro B, Guarasci F, Rinoldo C, Gambino C. Multiple productive cycles in the same year in soilless table grape cultivation. In: Proceedings of 35th Word Congr. Izmir, Turkey: Vine and Wine; 2012. p. 20

[8] Savvas D, Gianquinto GP, Tüzel Y, Gruda N. Soilless culture. In: Good Agricultural Practices for Greenhouse Vegetable Crops–Principles for Mediterranean Climate Areas. Plant Production and Protection Paper. Vol. 217. Rome: FAO; 2013. pp. 303-354 Available from: https://www.fao.org/3/i6787e/i6787e.pdf

[9] Sabır A, Sabir F, Yazar K, Kara Z. Investigation on development of some grapevine cultivars (V. vinifera L.) in soilless culture under controlled glasshouse condition. Current Trends in Technology and Science. 2016;**5**(3):622-626

[10] Baştaş PC, Tangolar S. The effect of growing medium and crop loads on yield and some quality properties of Prima grape cultivar grown in soilless culture. Alatarım. 2018;**17**(2):98-109 Available from: https://www.cabdirect.org/cabdirect/FullTextPDF/2020/20203216642.pdf

[11] Tangolar S, Tangolar S, Alkan Torun A, Tarım G, Ada M. Evaluation of table grape cultivation in soilless culture system. Türkiye Tarımsal Araştırmalar Dergisi. 2017;**4**(2):163-170

[12] Tangolar S, Tangolar S, Ada M, Tarım G, Alkan Torun A, Ertargın E. The effects of different nitrogen and potassium levels on yield and quality of two early grape cultivars grown in different soilless media. ISHS Acta Horticulturae. 2019;**1242**:349-356

[13] Kaya S, Tangolar S, Tangolar S. The effect of root pruning applications on

yield and quality of some table grape varieties grown in different solid culture media. Bahçe. 2018;**47**(Special issue 1: Türkiye 9. Bağcılık Teknolojileri Sempozyumu):575-585

[14] Atalan M. The Effects of Different Substrats and Chemical and Organic Fertilizer Applications on Yield and Quality in Early Cardinal Grape Grown in Soilless Culture [Msc thesis]. Gradute School of Natural and Applied Sciences. Adana, Türkiye: Department of Horticulture, Cukurova University; 2020. p. 107

[15] Zareei E, Zaare-Nahandi F, Hajilou J, Oustan S. Eliciting effects of magnetized solution on physiological and biochemical characteristics and elemental uptake in hydroponically grown grape (Vitis vinifera L. cv. Thompson Seedless). Plant Physiology and Biochemistry. 2021;**167**:586-595. DOI: 10.1016/j.plaphy.2021.08.036

[16] Tangolar S, Tangolar S, Alkan Torun A, Tarım G, Ada M, Aydın O. Effects of different nutrient solutions and growing media on yield and quality in grape growing in soilless culture. Mediterranean Agricultural Sciences. 2019;**32**:127-133

[17] Prieur C, Rigaud J, Cheynier V, Michel MM. Oligomeric and polymeric procyanidins from grape seeds (Vitis vinifera). Phytochemistry. 1994;**36**:781-784. DOI: 10.1016/S0031-9422(00)89817-9

[18] Ağaoğlu YS. Bilimsel ve Uygulamalı Bağcılık Cilt:1 Asma Biyolojisi. Kavaklıdere Eğitim Yayınları. 1. Ankara; 1999. p. 205

[19] Mironeasa S, Mironeasa C, Codină GG. Evaluation of mineral element content in grape seed and defatted grape seed. Food and Environment Safety-Journal of Faculty of Food Engineering, Ştefan cel Mare University-Suceava Year IX. 2010;**2**:53-60 Available from: http://fia-old.usv.ro/fiajournal/index.php/FENS/article/view/409/407

[20] Gök Tangolar S, Özoğul Y, Tangolar S, Torun A. Evaluation of fatty acid profiles and mineral content of grape seed oil of some grape genotypes. International Journal of Food Sciences and Nutrition. 2009;**60**:32-39. DOI: 10.1080/09637480701581551

[21] Gök Tangolar S, Özoğul F, Tangolar S, Yağmur C. Tocopherol content in fifteen grape varieties obtained using a rapid HPLC method. Journal of Food Composition and Analysis. 2011;**24**:481-486. DOI: 10.1016/j.jfca.2010.08.003

[22] Jordao AM, Correia AC. Relationship between antioxidant capacity, proanthocyanidin and anthocyanin content during grape maturation of Touriga Nacional and Tinta Roriz grape varieties. South African Journal of Enology and Viticulture. 2012;**33**:214-224. DOI: https://doi.org/10.21548/33-2-1121

[23] Jordao AM, Correia AC, Gonçalves FJ. Evolution of antioxidant capacity in seeds and skins during grape maturation and their association with proanthocyanidin and anthocyanin content. Vitis. 2012;**51**:137-139 Available from: https://citeseerx.ist.psu.edu/viewdoc/download?doi=10.1.1.940.9930&rep=rep1&type=pdf

[24] Ovcharova T, Zlatanov M, Dimitrova R. Chemical composition of seeds of four Bulgarian grape varieties. Ciência Técnica Vitivinicola. 2016a;**31**: 31-40. DOI: 10.1051/ctv/20163101031

[25] Tangolar S, Turan M, Tangolar S, Ateş F. Evaluation of amino acid contents and enzyme activities of seeds from

Semillon and Carignane wine grape cultivars grown under different irrigation conditions. Scientia Horticulturae. 2019;**251**:181-188. DOI: 10.1016/j.scienta.2019.02.079

[26] Gutierrez-Gamboa G, Garde-Cerdan T, Rubio-Breton P, Perez-Alvarez EP. Study of must and wine amino acid composition afetr seaweed applications to Tempranillo blanco grapevines. Food Chemistry. 2020;**308**:25605. DOI: 10.1016/j.foodchem.2019.125605

[27] Huskisson E, Maggini S, Ruf M. The influence of micronutrients on cognitive function and performance. Journal of International Medical Research. 2007;**35**(1):1-19. DOI: 10.1177/147323000703500101

[28] Key MN, Zwilling CE, Talukdar T, Barbey AK. Essential amino acids, vitamins, and minerals moderate the relationship between the right frontal pole and measures of memory. Molecular Nutrition & Food Research. 2019;**63**(15):1801048. DOI: 10.1002/mnfr.201801048

[29] Huang Z, Ough CS. Amino acid profiles of commercial grape juices and wines. American Journal of Enology and Viticulture. 1991;**42**:261-267 Available from: https://www.ajevonline.org/content/42/3/261.short

[30] Esteban MA, Villanueva MJ, Lissarrague JR. Effect of Irrigation on changes in berry composition of Tempranillo during maturation. Sugars, organic acids, and mineral elements. American Journal of Enology and Viticulture. 1999;**50**:418-434 Available from: https://www.ajevonline.org/content/50/4/418.short

[31] De La Hera-Orts M, Martinez-Cutillas A, Lopez-Roca JM, Gomez-Plaza E. Effect of moderate irrigation on grape composition during ripening. Spanish Journal of Agricultural Research. 2005;**3**:352-361

[32] Downey MO, Dokoozlian NK, Krstic MP. Cultural practice and environmental impacts on the flavonoid composition of grapes and wine: A review of recent research. American Journal of Enology and Viticulture. 2006;**57**:257-268 Available from: https://www.ajevonline.org/content/ajev/57/3/257.full.pdf

[33] Kurtural SK, Dami IE, Taylor BH. Effects of pruning and cluster thinning on yield and fruit composition of Chambourcin grapevines. HortTechnology. 2006;**16**:233-240. DOI: 10.21273/HORTTECH.16.2.0233

[34] Wessner LF, Kurtural SK. Pruning systems and canopy management practice interact on the yield and fruit composition of Syrah. American Journal of Enology and Viticulture. 2013;**64**: 134-138. DOI: 10.5344/ajev.2012.12056

[35] Abdrabba S, Hussein S. Chemical composition of pulp, seed and peel of red grape from Libya. Global Journal of Scientific Researches. 2015;**3**:6-11 Available from: https://docplayer.net/33310181-Chemical-composition-of-pulp-seed-and-peel-of-red-grape-from-libya.html

[36] Bouzas-Cid Y, Falqué E, Orriols I, Trigo-Córdoba E, Díaz-Lozada E, Fornos-Rivas D, et al. Amino acids profile of two Galician white grapevine cultivars (Godello and Treixadura). Ciência e Técnica Vitivinicola. 2015;**30**:84-93. DOI: 10.1051/ctv/20153002084

[37] Gök Tangolar S, Tangolar S, Tarım G, Kelebek H, Topçu S. The effects of bud load and applied water amounts on the biochemical composition of the 'Narince'

grape variety (Vitis vinifera L.). Notulae Botanicae Horti Agrobotanici. 2015;**43**:380-387. DOI: 10.15835/nbha4329958

[38] Gök Tangolar S, Tangolar S, Kelebek H, Topcu S. Determination of phenolics, sugars, organic acids and antioxidants in the grape variety Kalecik Karası under different bud loads and irrigation amounts. Horticultural Science & Technology. 2016;**34**(3):495-509. DOI: 10.12972/kjhst.20160050

[39] Hussein S, Abdrabba S. Physico-chemical characteristics, fatty acid, composition of grape seed oil and phenolic compounds of whole seeds, seeds and leaves of red grape in Libya. International Journal of Applied Science and Mathematics. 2015;**2**:2394-2894

[40] Nelson CC, Kennedy JA, Zhang Y, Kurtural SK. Applied water and rootstocks affect productivity and anthocyanin composition of Zinfandel in central California. American Journal of Enology and Viticulture. 2016;**67**:18-28. DOI: 10.5344/ajev.2015.15043

[41] Sánchez-Gómez R, Garde-Cerdán T, Zalacain A, Garcia R, Cabrita MJ, Salinas MR. Vine-shoot waste aqueous extract applied as foliar fertilizer to grapevines: Effect on amino acids and fermentative volatile content. Food Chemistry. 2016;**197**:132-140. DOI: 10.1016/j.foodchem.2015.10.034

[42] Gutiérrez-Gamboa G, Garde-Cerdán T, Gonzalo-Diago A, Moreno-Simunovic Y, Martínez-Gil AM. Effect of different foliar nitrogen applications on the must amino acids and glutathione composition in Cabernet Sauvignon vineyard. LWT-Food Science and Technology. 2017;**75**:147-154. DOI: 10.1016/j.lwt.2016.08.039

[43] Canoura C, Kelly MT, Ojeda H. Effect of irrigation and timing and type of nitrogen application on the biochemical composition of Vitis vinifera L. cv. Chardonnay and Syrah grape berries. Food Chemistry. 2018;**241**:171-181. DOI: 10.1016/j.foodchem. 2017.07.114

[44] Wu J, Drappier V, Hilbert G, Guillaumie S, Dai Z, Geny L, et al. The effects of a moderate grape temperature increase on berry secondary metabolites. Oeno One. 2019;**53**(2):321-333. DOI: 10.20870/oeno-one.2019.53.2.2434

[45] Gutiérrez-Gamboa G, Carrasco-Quiroz M, Martínez-Gil AM, Pérez-Álvarez EP, Garde-Cerdán T, Moreno-Simunovic Y. Grape and wine amino acid composition from Carignan noir grapevines growing under rainfed conditions in the Maule Valley, Chile: Effects of location and rootstock. Food Research International. 2018;**105**:344-352. DOI: 10.1016/j.foodres.2017.11.021

[46] Gutiérrez-Gamboa G, Portu J, López R, Santamaria P, Garde-Cerdán T. Elicitor and nitrogen applications to Garnacha Graciano and Tempranillo vines: Effect on grape amino acid composition. Journal of the Science of Food and Agriculture. 2018;**98**:2341-2349. DOI: 10.1002/jsfa.8725

[47] Bouzas-Cid Y, Díaz-Losada E, Trigo-Córdoba E, Falqué E, Orriols I, Garde-Cerdán T, et al. Effects of irrigation over three years on the amino acid composition of Albariño (Vitis vinifera L) musts and wines in two different terroirs. Scientia Horticulturae. 2018;**227**:313-325. DOI: 10.1016/j.scienta.2017.05.005

[48] Bouzas-Cid Y, Falqué E, Orriols I, Mirás-Avalos JM. Effects of irrigation over three years on the amino acid composition of Treixadura (Vitis vinifera L.) musts and wines, and on the aromatic composition and sensory profiles of its wines. Food Chemistry. 2018;**240**:707-716. DOI: 10.1016/j.foodchem.2017.08.013

[49] Bouzas-Cid Y, Trigo-Córdoba E, Falqué E, Orriols I, Mirás-Avalos JM. Influence of supplementary irrigation on the amino acid and volatile composition of Godello wines from the Ribeiro Designation of Origin. Food Research International. 2018c;**111**:715-723. DOI: 10.1016/j.foodres.2018.05.074

[50] Bremner JM. Total nitrogen. In: Black CA, editor. Methods of Soil Analysis. Part 2: Chemical and Microbial Properties. Number 9 in series Agronomy. Madison, USA: American Society of Agronomy, Inc. Publisher; 1965. pp. 1049-1178

[51] Kacar B. Bitki ve Toprağın Kimyasal Analizleri. II. Bitki Analizleri. Ankara Üniversitesi Ziraat Fakültesi Yayınları 453, Uygulama Kılavuzu 155. Ankara: Ankara Üniversitesi Basımevi; 1972

[52] Antoine FR, Wie CI, Littell RC, Marshall MR. HPLC method for analysis of free amino acids in fish using o-phthaldialdehyde precolumn derivatization. Journal of Agricultural and Food Chemistry. 1999;**47**:5100-5107

[53] Kitir N, Gunes A, Turan M, Yildirim E, Topcuoglu B, Turker M, et al. Bio-boron fertilizer applications affect amino acid and organic acid content and physiological properties of strawberry plant. Erwerbs-obstbau. 2019;**61**(2):129-137. DOI: 10.1007/s10341-018-0409-3

[54] Bates LS. Rapid determination of free proline for water-stress studies. Plant and Soil. 1973;**39**:205-207 Available from: https://link.springer.com/content/pdf/10.1007/BF00018060.pdf

[55] Sahin A, Kıran Y, Karatas F, Sonmez S, Vitamins A. C and E and β-carotene content in seeds of seven species of Vicia L. Journal of Integrative Plant Biology. 2005;**47**:487-493. DOI: 10.1111/j.1744-7909.2005.00083.x

[56] Al-Saleh IA, Billedo G, El-Doush II. Levels of selenium DL α-tocopherol, DL- γ-tocopherol, all-trans-retinol, thymoquinone and thymol in different brands of Nigella sativa seeds. Journal of Food Composition and Analysis. 2006;**19**:167-175. DOI: 10.1016/j.jfca.2005.04.011

[57] Mozumder NHMR, Akhter MJ, Khatun AA, Rokibuzzaman M, Akhtaruzzaman M. Estimation of water-soluble vitamin B-complex in selected leafy and non-leafy vegetables by HPLC method. Oriental Journal of Chemistry. 2019;**35**:1344-1351. DOI: 10.13005/ojc/350414

[58] AOAC. Official Methods of Analysis of the Association of Official Analytical Chemists. 15th ed. Arlington, VA: Association of Official Analytical Chemists; 1990. pp. 1058-1059

[59] Král M, Tauferová A, Tremlová B, Šnirc M, Árvay J, Walczycka M, et al. Macro-and micro-elements in locally produced and imported fruits on Czech market: A quantitative assessment. Erwerbs-obstbau. 2020;**62**(3):361-367

[60] Cantürk S, Kunter B, Keskin N. Gülüzümü (Vitis vinifera L.) çeşidinin fitokimyasal özellikleri üzerinde araştırmalar. Selcuk Journal of Agriculture and Food Sciences. 2013;**27**:359-364 Available from: http://sjafs.selcuk.edu.tr/sjafs/article/view/43

[61] Anonymous. FoodData Central Search Results. Grapes, Red or Green (European Type, such as Thompson Seedless), Raw [Internet]. 2019. Available from: https://fdc.nal.usda.gov/fdc-app.html#/food-details/174683/nutrients [Accessed: December 10, 2021]

[62] Olsen N, Ware M. What are the Health Benefits of Grapes? [Internet]. 2017. Available from:

https://www.medicalnewstoday.com/articles/271156#nutrition. [Accessed: December 10, 2021].

[63] Bourre JM. Effects of nutrients (in food) on the structure and function of the nervous system: Update on dietary requirements for brain. Part 1: Micronutrients. Journal of Nutrition Health and Aging. 2006;**10**(5):377 Available from: http://citeseerx.ist.psu.edu/viewdoc/download?doi=10.1.1.455.3456&rep=rep1&type=pdf

[64] Cetin ES, Altinöz D, Tarçan E, Baydar NG. Chemical composition of grape canes. Industrial Crops and Products. 2011;**34**(1):994-998. DOI: 10.1016/j.indcrop. 2011. 03.004

[65] Whelton PK, He J, Cutler JA, Brancati FL, Appel LJ, Follmann D, et al. Effects of oral potassium on blood pressure: Meta-analysis of randomized controlled clinical trials. Journal of the American Medical Association. 1997;**277**(20):1624-1632. DOI: 10.1001/jama.1997. 03540440058033

[66] He FJ, MacGregor GA. Beneficial effects of potassium on human health. Physiologia Plantarum. 2008;**133**(4):725-735. DOI: https://doi.org/10.1111/j.1399-3054.2007.01033.x

[67] Brody T. Nutritional Biochemistry. San Diego: Academic; 1994

[68] Hemalatha S, Platel K, Srinivasan K. Zinc and iron contents and their bioaccessibility in cereals and pulses consumed in India. Food Chemistry. 2007;**102**(4):1328-1336. DOI: 10.1016/j.foodchem.2006.07.015

[69] Talwar GP, Srivastava LM, Mudgil KD. Textbook of Biochemistry and Human Biology, 2nd ed. Prentice Hall of India Private Ltd, India. 1989 Available from: https://www.amazon.com/Textbook-Biochemistry-Human-Biology-Talwar/dp/8120319656

[70] Schachter M. The Importance of Magnesium to Human Nutrition. 1996. Available from: http://www.outsideinlife.com/upload/PDF/magnesium.pdf

[71] Bohnert HJ, Jensen RG. Strategies for engineering water- stress tolerance in plants. Trends in Biotechnology 1996;14:89-97. Available from: https://reader.elsevier.com/reader/sd/pii/0167779996809292?token= 4C87E30A51717D67AE7DCED16DF40A5ACBEB4AF38C886B5E4AF4F5748DA05E9E70BC1160131B059C25DA017921B25399&originRegion=eu-west-1&originCreation=20211217145024

[72] Anjum SA, Xie X, Wang L, Saleem MF, Man C, Lei W. Morphological, physiological and biochemical responses of plants to drought stress. African Journal of Agricultural Research. 2011;**6**:2026-2032. DOI: 10.5897/AJAR10.027

[73] Liang X, Zhang L, Natarajan SK, Becker DF. Proline mechanism of stress survival. Antioxids & Redox Signal. 2013;**19**:998-1011. DOI: 10.1089/ars.2012.5074

[74] Arabshahi M, Mobasser HR. Effect of drought stress on carotenoid and chlorophyll contents and osmolyte accumulation. Medicinal Chemistry Research. 2017;**2**:193-197

[75] Fernandez-Novales J, Garde-Cerdan T, Tardaguila J, Gutierrez-Gamboa G, Perez-Alvarez EP, Diago MP. Assesment of amino acids and total soluble solid in intact grape berries using contactless VIS and NIR spectroscopy during ripening. Talanta. 2019;**199**:244-253. DOI: 10.1016/j.talanta.2019.02.037

[76] Valdés ME, Talaverano MI, Moreno D, Prieto MH, Mancha LA, Uriarte D, et al. Effect of the timing of water deficit on the must amino acid profile of Tempranillo grapes grown under the semiarid conditions of SW Spain. Food Chemistry. 2019;**292**:24-31. DOI: 10.1016/j.foodchem.2019.04.046

Chapter 2

Nutrient Solution for Hydroponics

Moaed Ali Al Meselmani

Abstract

Hydroponics is a profitable, sustainable agricultural method and environmental friendly technology for growing plants without soil. It is the fastest-growing agriculture sector, rapidly gaining momentum and popularity, and could dominate food production in the future. Nutrient solution and its management are the cornerstone of a successful hydroponic system and are the most important determinant of crop production and quality, which is largely dependent on the extent to which plant nutrients are acquired from the nutrient solution. All nutrients in the solution in balanced ratio are supplied directly to the plants and the composition of the solution must reflect the uptake ratio of individual elements by the crop. A balanced supply of nutrients is a prerequisite for the efficient use of resources, and stabilization of the solution pH, electrical conductivity, O_2 level, and temperature is essential for optimum crop yield in hydroponic systems. In this chapter, the composition of the nutrient solution, nutrient availability which is affected by many factors, and the management of the nutrient solution are discussed.

Keywords: nutrient solution, pH, EC, solution temperature, nutrient management

1. Introduction

Hydroponic techniques have been developed to facilitate cultivation under diverse environments and to improve farming practices using soilless methods. In this novel world, hydroponic farming makes efficient use of fertilizers and water, increases productivity, and provides better crop quality; **Table 1** shows the difference in productivity between soil and soilless culture for different crops [1]. Also, due to the risks of soil and water contamination in metropolitan areas, this technique has a potential alternative to agricultural production in cities. Hydroponic systems irrespective of their scale reduce dependence on the soil as a substrate and instead derive nutrition directly from the hydroponic solution comprising of water and nutrients [2]. Because hydroponics provides better control of plant growth, it is possible to achieve high quality and productivity through careful management of—nutrient composition, dissolved O_2 concentration, temperature, pH, and electrical conductivity (EC) of the nutrient solution. Nutrient supply in hydroponics can significantly influence the nutrition, taste, texture, color, and other characteristics of fruit and vegetable crops [3]. In hydroponics, essential nutrient elements are dissolved in appropriate concentrations and relative ratios to achieve the normal growth of plants [4]. It is well known that the productivity and quality of crops grown in hydroponic systems are markedly dependent on the extent of the plant nutrients acquisition

IntechOpen

Crop	Hydroponic system production (Kg/ha)	Open-field production (Kg/ha)
Rice	13,456.56	841.03–1009.25
Maize	8971.0	1682.07
Peas	15,699.32	2242.76
Tomato	403,335.81	11,203.75–22,407.47
Potato	156,852.29	17,925.98
Cabbage	20,184.84	14,577.94
Cucumber	31,398.64	7849.66
Lettuce	23,548.98	10.092.42

Source: Singh and Singh [1].

Table 1.
Yield comparisons for different crops between hydroponic and open field cultivation.

from the growing medium [5]. Due to this, nutrient solution and its management are the cornerstone for a successful hydroponics system and are the most important determining factors of crop yield and quality.

2. Plant nutrients

Plant nutrients used in hydroponics are dissolved in water and are mostly in inorganic and ionic forms. All the essential elements for plant growth are supplied using different chemical combinations and establishing a nutrient solution that provides a favorable ratio of ions for plant growth and development is considered an important step in cultivating crops in hydroponic systems [6]. Plant uptake of nutrients can only proceed when they are present in an available form for absorption, and in most situations, nutrients are absorbed in an ionic form. Ions are electrically charged forms of each nutrient, some are cations (positively charged) and others are anions (negatively charged). For example, nitrogen is absorbed as ammonium (NH_4^+, a cation) or nitrate (NO_3^-, an anion); **Table 2** shows the available form of each nutrient and different nutrient solution formulas which have been established by many scientists. There are various standard nutrient solutions, such as the Hoagland and Snyder [13], Hoagland and Arnon [11], Steiner [14] Bollard [15], and others. These standard solutions are good as a general guideline but are not adapted to specific growing conditions. The function of a hydroponics nutrient solution is to supply the plant roots with water, oxygen, and essential mineral elements in soluble form. A nutrient solution usually contains inorganic ions from soluble salts of essential elements required by the plant. However, some organic compounds such as iron chelates may be present [16]. A total of 17 elements are considered essential for most plants, these are carbon, hydrogen, oxygen, nitrogen, phosphorus, potassium, calcium, magnesium, sulfur, iron, copper, zinc, manganese, molybdenum, boron, chlorine, and nickel [17]. An essential element has a clear physiological role, and its absence prevents the complete plant life cycle [18]. Among the minerals, N, P, and K are the most decisive elements in plants [6]. Some other elements such as sodium, silicon, vanadium, can stimulate growth, or can compensate for the toxic effects of other elements, or may replace essential nutrients in a less specific role. Tahereh et al. [19] reported that the plants grown in the absence

Element	Form taken up by plants	Hoagland & Arnon	Hewitt	Cooper	Steiner
		mg L^{-1}			
Nitrogen	NH_4^+, NO_3^-	210	168	200–236	168
Phosphorus	HPO_4^{-2}, $H_2PO_4^-$	31	41	60	31
Potassium	K^+	234	156	300	273
Calcium	Ca^{2+}	160	160	170–185	180
Magnesium	Mg^{2+}	34	36	50	48
sulfur	SO_4^{2-}	64	48	68	336
Iron	Fe^{2+}, Fe^{3+}	2.5	2.8	12	2–4
Copper	Cu^{2+}	0.02	0.064	0.1	0.02
Zinc	Zn^{2+}	0.05	0.065	0.1	0.11
Manganese	Mn^{2+}, Mn^{4+}	0.5	0.54	2	0.62
Boron	H_3BO_3, BO_3^-, $B_4O_7^{2-}$	0.5	0.54	0.3	0.14
Molybdenum	MoO_4^{2-}	0.01	0.04	0.2	Not considered

Source: Salisbury and Ross [7]; Cooper [8]; Steiner [9]; Windsor and Schwarz [10]; Hoagland and Arnon [11]; Hewitt [12].

Table 2.
Nutrient's form taken up by plants and nutrients compositions as suggested by different scientists.

of silica would be weak and show abnormal growth, and proper application of this nutrient can increase consistency and disease resistance, reduce the outbreak of nutrient deficiencies, improve product quality and increase crop yield. In hydroponics, all the nutrients are in a balanced ratio which is directly supplied to the plants, and composition must reflect the uptake ratio of individual elements by the crop, as the demand between species differs, and must be specific for each crop [20]. It is very important to keep ionic balance in the nutrient solution since plant growth and productivity can be negatively affected by the improper relationship between the essential nutrients, that is, the ratio of anions: NO_3^-, $H_2PO_4^-$ and SO_4^{2-}, and the cations K^+, Ca^{2+}, Mg^{2+} [21], and a change in the concentration of one ion must be accompanied by either a corresponding change for an ion of the opposite charge, a complementary change for other ions of the same charge, or both [12]. However, for most common crop plants, critical levels for most nutrients have been determined [22].

2.1 Plant nutrients interaction

Nutrients in the nutrient solution have great interactions that may gain either positive or negative effects on crop production, depending on crop growth stages, amounts, combinations, and balance [23]. Inadequate or excessive concentrations of minerals or an imbalanced ion composition in the nutrient solution may inhibit plant development, resulting in toxicity or nutrient-induced deficiencies [24]. In crop plants, the nutrient interactions are generally measured in terms of growth response and change in concentration of nutrients. Nutrient interactions may be positive or negative and also possible to have no interactions. Interaction between nutrients occurs when the supply of one nutrient affects the absorption and utilization of other nutrients. This type of interaction is most common when one nutrient is in excess concentration in the growth medium. Upon the addition of two nutrients, an increase in crop yield that is more than adding only one, the interaction is positive (synergistic). Similarly, if adding the two nutrients together produced less yield as compared to individual ones, the interactions are negative (antagonistic). When there is no change, there is no interaction. However, most interactions are complex and better understanding of nutrient interactions may be useful in understanding the importance of a balanced supply of nutrients and consequently improvement in plant growth or yields [25]. According to Marschner [26], at the level of the nutrient acquisition mechanisms, competitive or antagonistic phenomena among elements can occur, for example, the interaction between NH_4^+ and K^+, and this could be crucial for NH_4^+ fed plants when exposed to a suboptimal/unbalanced availability of K^+ because the competition could induce/exacerbate K^+ deficiency [27], and it is more relevant when the additional application of NH_4^+ is of pivotal role to achieve specific qualitative objectives of the edible fruits [28]. The interactions between K^+/Na^+ and Cl^-/NO_3^- could represent a limiting factor for soilless cultivation of crop plants, especially in a semiarid environment characterized by saline water. NaCl interferes with the uptake processes of both K^+ and NO_3^-, since K^+ is sensitive to Na^+ in the external environment, while the uptake of NO_3^- is inhibited by Cl^- [29]. This phenomenon could be even more pronounced in hydroponic solutions particularly when used in a closed system, where monitoring the ratio between Ca^{2+}, Mg^{2+}, and K^+ in the solutions is very important to avoid K^+/Ca^{2+} induced Mg^{2+} deficiency. Calcium, magnesium, and potassium compete with each other and the addition of any one of them will reduce the uptake rate of the other two [26]. Unbalanced fertilization practice, with a high level of K^+ and Ca^{2+}, can induce Mg^{2+} deficiency in crop plants, Schimansky [30] suggested that the excessive availability of K^+ and Ca^{2+} could inhibit Mg^{2+} uptake by roots. Similarly, very high rates of Mg^{2+} fertilizers will depress K^+ absorption by plants, but this antagonism is not nearly as strong as the inverse relation of K^+ on Mg^{2+} [31]. Also, the uptake of nitrogen, sulfur, and iron is not exclusively dependent on its availability in the hydroponic solution but also on the presence of other elements. The uptake of NO_3^- was hampered by the shortages of iron and sulfur, and the effect on the assimilation process seems to play a dominant role in determining the NO_3^- accumulation at the leaf level. In the case of nitrogen and sulfur, the lacking one represses the assimilation of the other and induces physiological changes aiming at re-balancing the contents in the plant [32]. One of the greatest issues concerning hydroponic productions is sulfur starvation due to a consistent accumulation of NO_3^- in plant leaves [33]. The anion which is taken up relatively slowly can also reduce the uptake speed of its counter-ion, as observed for SO_4^{2-} on K^+ uptake [26].

In hydroponic solutions, interactions among solutes cannot be neglected and therefore ion activity should be used in calculations instead of concentrations [34]. The high ionic concentrations can disrupt membrane integrity and function, as well as internal solute balance and nutrient absorption, resulting in nutritional deficiency symptoms similar to those observed when nutrient concentrations are below the required levels [24]. In addition, the root physiological process is not only affected by the availability levels of the nutrients, but also by the nutrient sources and/or by the interactions among the different nutrients [35]. The chemical forms of a nutrient are also very important, for example, plants can use a wide variety of nitrogen forms, ranging from the inorganic, namely NH_4^+ and NO_3^-, to the organic ones, like urea and amino acids [36]. Ammonium is an attractive nitrogen form for root uptake due to its permanent availability and the reduced state of the nitrogen; nevertheless, when both nitrogen forms are supplied to the nutrient solution, plant roots may absorb preferentially one of them, depending on the heredity of each specie [37]. Pure NH_4^+ nutrition caused the development of toxicity symptoms in many herbaceous plants, as well as inhibited NO_3^- uptake [38]. Therefore, a balanced nitrogen diet is clearly beneficial for several plant species as compared to that based exclusively on NO_3^- [39]. Tomato root growth was optimal when NO_3^- and NH_4^+ were supplied in a 3:1 ratio; on the contrary, when NH_4^+ concentration was too high, a strong inhibition in the root development was observed [40]. However, the form of nitrogen suitable for obtaining the maximum production for each species and its cultivation conditions has not yet been defined [37]. Also, the plant species and environmental conditions are two critical factors that affect the optimum NO_3^-/NH_4^+ ratio, which can affect not only root development and morphology but also the overall root biomass. According to [41], the chemical quality of nutrient solutions can affect plant yield and bioactive compounds.

Several physical-chemical phenomena can alter the nutrient availability for plants, the most important of which are—precipitation, co-precipitation, and complexation. Precipitation reactions may occur when cations and anions in an aqueous solution combine to form a precipitate. It is known that phosphate availability can be reduced at pH above 7 mostly due to precipitation with calcium and different calcium-phosphate minerals can potentially form above this pH [42], and precipitation of phosphates must be avoided in hydroponic solutions because it is not only depleting phosphorus from the nutrient solution, but it may also reduce the solubility of other nutrients, such as calcium, magnesium, iron, and manganese. Also, sulfur availability can be limited by precipitation with calcium, as calcium-sulfate minerals [43]. Co-precipitation also may strongly reduce the solubility of nutrients added at trace concentrations, such as copper, zinc, manganese, and nickel, when insoluble compounds, such as iron hydroxides, calcium carbonates, or calcium phosphates, are formed [44]. In hydroponic solutions, a complex chemical compound is formed when a metal nutrient is bound by one or more neutral molecules or anions, either of organic or inorganic nature. The resulting complex can be a neutral compound, a cation, or an anion, depending on whether positive or negative charges prevail. These reactions diminish the concentration of the free ions in the nutrient solution, changing elemental bioavailability. The addition, organic ligands, such as: ethylenediaminetetraacetic acid (EDTA), Diethylenetriamine Penta acetic Acid (DTPA), Ethylenediamine (O-Hydroxyphenyl acetic) Acid (EDDHA), and citrate, can increase the stability of certain elements in solution, especially iron, copper, and zinc [45].

3. PH level of the nutrient solution

The pH value of the nutrient solution greatly affects plants' growth. This is because the nutrients added to the nutrient solution are available for the uptake by the plant are soluble in water only at particular pH levels, as shown in **Figure 1**. According to Mayavan et al. [47], the plants require a range of pH values to be maintained to ensure the availability of all the nutrients for uptake by the plants. Nutrient solution pH is typically managed between 5.5 and 6.5, and it seems to be a range where almost all hydroponically grown crops exhibit normal growth and nutrient uptake, and the optimum pH range for different crops grown hydroponically are shown in **Table 3**. However, species-specific pH responses of leafy greens grown in liquid culture hydroponic systems are largely unexplored [49]. However, the optimum pH for maximum growth differs not only between species, but also between cultivar,

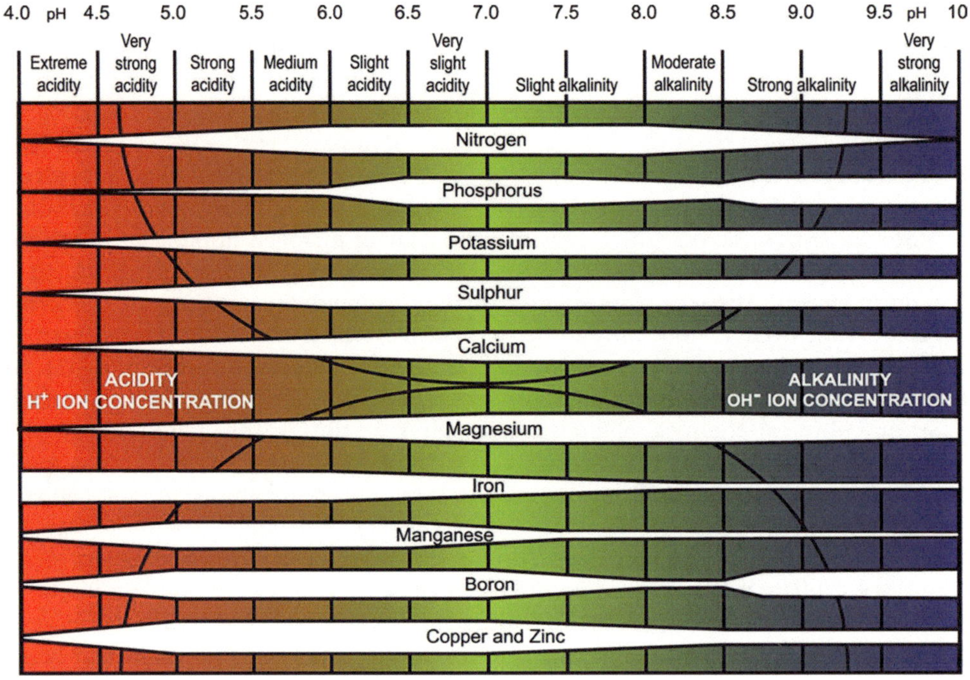

Figure 1.
The availability of different nutrients at the different pH bands is indicated by the width of the white bar: The wider the bar, the more available is the nutrient. Source: Truog [46].

Crop	Optimum pH range
Tomato	6.0–6.5
Pepper	5.5–6.0
Egg plant	6.0
Cucumber	5.0–5.5
Strawberry	6.0
Courgettes	6.0
Banana	5.5–6.5
Ficus	5.5–6.0
Spinach	6.0–7.0
Lettuce	6.0–7.0
Cabbage	6.5–7.0
Broccoli	6.0–6.8
Asparagus	6.0–8.0
Bean	6.0
Basil	5.5–6.0
Sage	5.5–6.5

Source: Sharma et al. [48].

Table 3.
The optimum range of pH values for different crops grown hydroponically.

climatic conditions, and soil, substrate, or nutrient solution conditions [50]. Frick and Mitchell [51] indicated that the pH of a hydroponic nutrient solution fluctuates because of the unbalanced anion and cation exchange reaction with roots and there is no buffering capacity in hydroponics. The changes in the pH of a nutrient solution depend on the difference in the magnitude of nutrient uptake by plants, in terms of the balance of anions over cations. When the anions are up taken in higher concentrations than cations, for example, nitrate, the plant excretes OH^- or HCO_3^- anions, to balance the electrical charges inside, which produces increasing in the pH value and this process is called physiological alkalinity [52]. Nutrient disorders and thereby growth reduction occur when pH is outside the optimum range, and several studies suggested that the direct effect of pH seems to be detrimental only at the extreme ends of acidity and alkalinity, and growth reductions and nutrient disorders outside of the conventional pH ranges can typically be attributed to pH-dependent factors [49, 53]. The growth response to pH is species-specific and further studies to investigate responses to pH of commercially important cultivars and species grown hydroponically need to be done [49]. In general, the pH of the plant root environment is affecting nutrient uptake, nutrient availability, ion antagonism, ionic species present, and solubility of fertilizer salts [50, 54]. Due to this, it is important to measure and maintain the pH value to the required level because a little drift in the pH value can make a lot of nutrients unavailable for the plants [47].

Precipitation/dissolution phenomena are often promoted by pH changes and, therefore, pH must be continuously controlled or buffered. Cations may form insoluble hydroxides at alkaline pH or other insoluble precipitates by reacting with other anionic nutrients. PH values above 7 may cause the precipitation of iron, zinc, copper, nickel, and manganese as insoluble hydroxides [55]. Also, at high pH values and high dissolved CO_2 concentrations, macronutrients like calcium and magnesium can precipitate as carbonates. As the pH increases above 7, most of the dissolved phosphorus reacts with calcium forming calcium phosphates. Gradually, reactions occur in which the dissolved free phosphate species form insoluble compounds that cause phosphate to become unavailable [56]. According to Resh [57], slightly acidic pH is optimum for hydroponic production because iron, manganese, calcium, and magnesium may form precipitates and become unavailable at pH above 7. Bugbee [58] also reported that availability of potassium and phosphorus is slightly reduced in a nutrient solution with high pH. The reason for the reduction in phosphorus uptake at a high pH level is explained by the reduction in the concentration of $H_2PO_4^-$, which is the substrate of the proton-coupled phosphate symporter in the plasma membrane, in the pH range of 5.6–8.5; conversely, a decrease in pH can increase the activity of proton-coupled solute transporters and enhance anion uptake [59]. Because pH affects nutrient availability and nutrient uptake across the plasma membrane, it is difficult to determine whether growth inhibition and nutrient disorders observed at low pH of the nutrient solution are a result of the direct effect of excessive hydronium ion concentration or pH-dependent factors affecting nutrient availability and uptake. At acidic pH, for example, in uncontrolled hydroponic systems under anoxic conditions, some elements might also precipitate as insoluble sulfides. Also, it is very important to note that, the addition of nutrients in the form of salts to hydroponic solutions may lead to hydrolysis reactions, which may result in the acidification or alkalinization of the medium. For example, nitrogen supply may alter solution pH, if nitrogen is added only in the form of NO_3^- (alkalinization) or NH_4^+ (acidification) [60].

In general, stabilizing the pH of a nutrient solution is necessary for optimum crop productivity in hydroponics [51], and maintaining an adequate nutrient solution and pH level are often cited as major obstacles to hydroponic production [61]. Despite the fact that the optimal pH in the root zone of most crops grown hydroponically ranges from 5.5 to 6.5, although values as low as 4.0 have been proposed for preventing the incidence of infections from Pythium and Phytophthora spp. [13, 49]. Low pH in the rhizosphere poses abiotic stress, resulting directly (i.e., high H^+ injury of roots) or/and indirectly (i.e., limited availability of phosphorus) in restricted plant growth and crop yield. The value of pH changes as the plant absorbs nutrients from the solution, the plants give hydrogen ions into the nutrients in exchange for the ions of elements they require, and they do this to be electrically neutral. The hydrogen ions that the plants get are a result of photosynthesis. These hydrogen ions combine with water to produce hydronium ions which increases the pH of the water. This has to be counteracted by adding acids like phosphoric acid into the nutrient solution to ensure the solubility of all the elements in the nutrients [47]. Various acids or bases used to adjust pH may also provide some interacting factors on the plants. For example, potassium hydroxide, sodium hydroxide, phosphoric acid, and acetic acid are commonly used to maintain the pH of the nutrient solution. The presence of these acids or bases may have had small impacts through the addition of minerals such as potassium, phosphorus, and/or sodium and the increased concentration of acetates. Other nonmineral nutrients containing acids (carbonic, formic, citric, acetylsalicylic, etc.) could be used for pH adjustment, but their potential toxicity and interactions with the nutrient solutions would need careful consideration and study. Overall, it would be ideal to have a solution where pH could be maintained easily within a small pH range without the addition of mineral nutrients [62]. Wang et al. [63] found that a mixture of three (HNO_3, H_3PO_4, and H_2SO_4) acids was much more effective than only single acid for maintaining an optimal solution pH of 5.5–6.5. The management of nutrient solution pH is an important challenge in soilless systems, since not only it may determine plant growth but also it influences dry matter production, root rhizosphere, and apoplastic pH [13]. However, in soilless culture, when maintaining marginal values of the optimum pH range, the risk of exceeding or dropping below them for some time increases due to the limited volume of nutrient solution per plant that is available in the root zone, and most plants, when exposed to external pH levels >7 or < 5, show growth restrictions. When soilless substrates are used instead of liquid-based hydroponics, pH in the nutrient solution interacts with substrates [64], and micronutrient toxicity occurs rather than deficiency. Therefore, the evaluation of the plant's pH response must consider the growing systems employed.

4. Nutrient solution electrical conductivity

In soilless culture, the total salt concentration of a nutrient solution must be considered, and the nutrient solution EC is an index of salt concentration and an indicator of electrolyte concentration of the solution and is related to the number of ions available to plants in the root zone. The EC is a measure of the total salts dissolved in the hydroponic nutrient solution. It is used for monitoring applications of fertilizers. However, EC reading does not provide information regarding the exact mineral content of the nutrient solution. It is an important factor that reflects the total content of macro- and micro-elements available to plants [6], and it is an easy and accurate method of measuring total salt concentration. Inadequate management of the nutrient solution, such as the use of a too high or a too low concentration of the nutrient solution, or an imbalanced ion composition could inhibit plant growth due to either toxicity or nutrient-induced deficiency [65]. In hydroponic production systems, EC management is one of the most important and manageable cultural practices that affects the visual, nutritional, and phytochemical quality of leafy vegetables [4]. However, managing the EC in moderately high levels—either by using low-quality water that contains residual ions, such as Cl^-, Na^+ and SO_4^-, or by adding major nutrients through stock solutions—is a cultivation management technique that provides great potential to achieve high dietary and organoleptic quality in fresh vegetables [24]. Each plant species has a proper uptake rate of the nutrient solution; excessively high or low levels of the nutrient solution have a negative effect on plants. For many leafy vegetables, there are already specific formulations used on a commercial scale for hydroponics, and the optimum EC levels for different crops grown hydroponically are shown in **Table 4**. Although the plants were supplied with suitable ion ratios, plants can easily suffer from nutrient deficiency or excess if the nutrient solution concentration is low or high. Therefore, it is crucial to determine the suitable EC level of nutrient solutions with favorable ion ratios for growing plants [6]. The optimal EC is crop specific and depends on environmental conditions [66]. Thus, the determination of the most favorable nutrient ratio for each species under diverse climatic conditions is of major importance.

Many studies have reported that EC levels of nutrient solutions affect the growth of various crops. The optimal EC level range should be from 1.5 to 3.5 dS m^{-1} for most hydroponic crops, but this value varies between crop species and phenological stages [6]. However, the upper levels of EC in nutrient solutions must be considered for each species, since excessive EC values may decrease the osmotic potential of the nutrient solution and consequently result in delays in water transport from roots to fruits, with negative effects on fruit expansion and yield [24]. The EC levels showed a considerable

Crop	EC (dSm^{-1})
Tomato	2.0–4.0
Pepper	0.8–1.8
Egg plant	2.5–3.5
Cucumber	1.7–2.0
Strawberry	1.8–2.2
Courgettes	1.8–2.4
Banana	1.8–2.2
Ficus	1.6–2.4
Spinach	1.8–2.3
Lettuce	1.2–1.8
Cabbage	2.5–3.0
Broccoli	2.8–3.5
Asparagus	1.4–1.8
Bean	2.0–4.0
Basil	1.0–1.6
Sage	1.0–1.6

Source: Sharma et al. [48].

Table 4.
Optimum range of EC values for different crops grown hydroponically.

influence on the ratio of ions as well as the uptake content of individual minerals. Too low and too high EC would reduce yields, visual quality, phytochemical compounds and lead to a less attractive color and taste to consumers, and enhance the negative health effects due to nitrate accumulation [4]. Increasing conductivity in nutrient solution may reduce water absorption by plants and decrease photosynthesis [67]. Also, higher EC means plants are exposed to salinity stress and high levels of nutrients, which hinders nutrient uptake and induces osmotic stress, ion toxicity, nutrient imbalance, wastes nutrients, and increases the discharge of more nutrients into the environment, resulting in environmental pollution. At the extreme EC level, plants are not able to take up any more water, and water will move backward out of the nutrient solution, which makes plants withered. The elevated EC may have negative effects on yield but can also positively affect the quality of the fresh produce, thus compromising any yield losses through the production of products with a high added value [24]. As an example, the yield of tomatoes under the hydroponic system increased as EC of the nutrient solution increased from 0 to 3 dS m^{-1} and decreased as the EC increased from 3 to 5 dS m^{-1} due to an increase in water stress [68]. Lower EC values mean the supply of some nutrients to the crop may be inadequate are mostly accompanied by nutrient deficiencies and decreasing yield [69]. So, appropriate management of EC in hydroponics technique can give an effective tool for improving vegetable yield and quality [48].

EC is modified by plants as they absorb nutrients and water from the nutrient solution. When a nutrient solution is applied continuously, plants can uptake ions at very low concentrations, and a high proportion of the nutrients are not used by plants. However, in particular situations, too low concentrations do not cover the minimum demand for certain nutrients. On the other hand, high concentrated nutrient solutions lead to excessive nutrient uptake and therefore toxic effects may be expected. Therefore, a decrease in the concentration of some ions and an increase in the concentration of others is observed simultaneously, both in close and open systems. It was observed, in a closed hydroponic system with a rose crop, that the concentration of iron decreased very fast, while that of Ca^{2+}, Mg^{2+}, and Cl^- increased; moreover, concentrations of K^+, Ca^{2+}, and SO_4^{2-} did not reach critical levels [70]. Providing the most suitable nutrient solution and EC level for growing vegetables and crops in hydroponic systems helps to avoid the waste of nutrient solution, which contributes to saving production costs for growing crops in plant factories and preventing environmental pollution, and the value of EC is required to be controlled to ensure nutritional elements needed by plants is fulfilled.

5. Nutrient solution temperature

Nutrient solution temperature is considered as one of the most important determining factors of crop yield and quality in hydroponic production systems [71]. The temperature of the nutrient solution affects the physiological process in the root, such as the absorption of water and nutrients, and the thermal regulation of hydroponic solution can contribute to improving and optimizing plant physiological processes [72]. Nutrient uptake for plants grown in glasshouses may be positively and adversely affected by manipulating the hydroponic solution temperature to the optimum level [73]. It is also possible that the increased temperature facilitated solubility of minerals and increase uptake since the rate of dissolving of solutes increases with increase in temperature [74], and the nutrient solution temperature tends to determine the concentration of nutrients absorbed by the plant, as more nutrients are dissolved at higher temperatures and less at lower temperatures, consequently influencing the efficiency of the photosynthetic apparatus [75]. Calatayud et al. [76] revealed that, in most plant species, nutrient uptake by roots decreased at low temperatures. Increasing nutrient temperatures increased nutrient uptake in cucumber and enhanced plant growth leading to a significant increase in yield [77]. The uptake rate of N, P, K, Na, Fe, Mn, and Zn in Jojoba was significantly reduced at low temperatures [78]. While, in cucumber, uptake of N, P, K, Ca, and Mg was increased when the temperature was raised in a closed hydroponic system from 12 to 20°C [77]. It has been reported that commercial growers experience a lower level of ornamental plant production in winter than in summer due to the low temperature of the solution [79, 80]. Also, the production of various plant metabolites is influenced by the temperature of the root zone in many plants, including leafy vegetables [67].

The chemical equilibrium of the solution is affected by nutrient temperature, and this is particularly crucial for areas where the over warming of the nutrient solution often occurs, impacting also all the physiological processes in the plant [81]. Generally, the cold solution increased NO_3^- uptake and thin-white roots production but decreased water uptake and it also influenced the photosynthetic apparatus. The temperature of the nutrient solution also has a direct relation to the amount of oxygen consumed by plants, and an inverse relation to the oxygen dissolved. It is of paramount importance to regulate hydroponic solution temperatures in situations whereby, plants are grown in a controlled environment during winter months. Optimizing solution temperature can be achieved by warming the nutrient solution and this showed success in a variety of crops [82, 83]. High temperature in the root zone is one of the most significant limiting factors for lettuce cultivation in tropical hydroponics. Instead of cooling the entire greenhouse air, the root zone cooling system could be an energy-efficient cooling system for a greenhouse for tropical hydroponics. Therefore, it is very important to study the optimum nutrient temperature requirements for different crops grown in climates with adverse winter conditions.

6. Dissolved oxygen levels in nutrient solution

Maintaining enough dissolved O_2 in a nutrient solution in a hydroponic system is crucial for plant health. Oxygen availability to roots grown in soilless culture can become limiting in case O_2 demand exceeds O_2 supply, inducing a reduction in root growth rate, ion, and water uptake, eventually reducing plant production [84]. Plants grown in hydroponic systems can quickly deplete the dissolved O_2 in the nutrient solution resulting in poor root aeration, especially when greenhouse temperatures are high, **Table 5** shows O_2 solubility in pure water at different temperatures. Jong et al. [86] noticed that cucumber growth was significantly affected by root-zone aeration. Roosta et al. [87] found improve eggplant growth with rising O_2 levels in the nutrient solution in floating hydroponic cultures and higher O_2 levels seemed to alleviate signs of ammonium toxicity among the tested plants. Root respiration also decreases when O_2 supply in the root environment falls below a critical O_2 concentration [88]. The sensitivity of roots to low O_2 concentration depends on its effect on mitochondrial respiration because it supplies most of the energy required for root function. Reduction in O_2 levels in the nutrient solution could lead to poor roots, an increase in the incidence of diseases and pests, and a reduction in plant growth. Oxygen around a plant's roots affects the beneficial microorganisms that provide protection from pathogens and improve nutrient uptake. Tomato plant roots would be

Temperature (°C)	Oxygen solubility (mg L^{-1})
10	11.29
15	10.08
20	9.09
25	8.26
30	7.56
35	6.95
40	6.41
45	5.93

Source: Trejo-Téllez et al. [85].

Table 5.
Solubility of oxygen in pure water at various temperatures.

much more susceptible to Pythium infection if root zone O_2 dropped below 2.8 mg/L [89]. Dissolved O_2 concentration, is strongly dependent on solution temperature and flow rate near the root zone, as well as on the growth rate of the crop, and may be influenced by the bacterial community present in the solution. The temperature has a direct relationship to the amount of oxygen consumed by the plant and a reverse relationship with dissolved oxygen from the nutrient solution. The consumption of O_2 increases when the temperature of the nutrient solution increases. Consequently, it produces an increase in the relative concentration of CO_2 in the root environment if the root aeration is not adequate [90]. For overcoming the limited oxygen exchange between the atmosphere and the nutrient solution in static deep water culture hydroponics, the nutrient solution is aerated by an air bubbler connecting with the pump to provide adequate root oxygenation [6]. Roots of loose-leaf lettuce grown in a floating raft hydroponic system were found to have a better condition with oxygen enrichment done in nutrient solution up to aeration pressure of 0.012 mPa and concentration of 600 ppm, with indicators of increasing length and total root surface area [68]. So, it is important to make sure the nutrient solution is properly aerated to maintain enough oxygen for the plant cells found in the root mass since this is crucial to the function of the plant's cells and the microbial world.

7. Nutrient solution management

An optimized and well-balanced supply of nutrients is a prerequisite for efficient use of the resources by hydroponically grown vegetables, not only to ensure a high yield but also to guarantee the quality of the edible tissues. In hydroponics, because of the limited nutrient-buffering capacity of the system and the ability to make rapid changes, careful monitoring of the system is necessary. The frequency and volume of the nutrient solution applied depends on the type of substrate, the crop and growth stage, the size of the container, the irrigation systems used, and the prevailing climatic conditions. Depending on the stage of plant development, some elements in the nutrient solution will be depleted more quickly than others and as water evaporates from the nutrient solution, the fertilizer becomes more concentrated and can burn plant roots. In hydroponics, nutrient management is very important and must be done as highly efficient as possible to improve productivity without harming the environment.

Nutrient management included- application the right fertilizers source (e.g., ammonium or nitrate as nitrogen source), balanced nutrient solution according to plants needs and according to plant growth stages and climatic conditions. The main principle of crop nutrient management is to prevent overapplication of nutrients, which prevents loss due to low yield from toxicities of some nutrients resulting from the unnecessary use of fertilizers. It was reported that the strong difference between the ion ratios presented in the nutrient solution and those absorbed by plants led to the accumulation of certain ions in the nutrient solution, which caused an imbalance of mineral elements in the nutrient solution and created more energy to absorb the suitable ions [6]. Recycling exhausted solutions may also represent an efficient strategy to prevent groundwater and environmental pollution. However, the main problem with the reuse of exhausted nutrient solutions is the shortage of some key macro and micronutrients [91] and their increased salinity [92] causing, in turn, problems for crops [93, 94]. Thus, it is very important to develop management practices/tools that reduce salinity in recycled solutions and/or minimize the physiological impact of salinity on plants. The salinity increase could be contrasted by treating the recycled water with appropriate osmotic systems, including forward and reverse osmosis.

In closed hydroponic systems, accumulation of potentially toxic organic compounds released by the roots of cultivated plants may occur and to overcome this issue, several treatment techniques have been proposed for root exudates degradation or removal. However, for the treatment to be effective, it should be able to remove root exudates without interfering with the inorganic mineral nutrients in the solution. As above-mentioned, the regulation of the solution flow rate in hydroponic production affects plant growth, which in turn affects crop yield and quality. The influence of nutrient solution flow rate on plant growth is related to the plants' physical environment. The flow of nutrient solution not only promotes nutrient ion diffusion but also increases the kinetic energy available to plant roots Therefore, adjusting the flow rate can improve plant yield and a reasonable flow pattern must be carefully selected. Because increasing the flow means increasing electricity consumption, it increases the cost of operation. Therefore, it is important to balance plant yield, nutrient management, and energy utilization. According to Baiyin et al. [95], determining the ideal flow rate for hydroponic production may help to increase yield. However, such a determination requires a specific analysis of each crop and growing environment. The hydroponic nutrient solution is the sole source of nutrients to the plant; therefore, it is imperative that a balanced solution, containing all the right plant nutrients, is applied.

8. Conclusion

Hydroponic cultivation is revolutionizing agricultural crop production techniques all over the world owing to its minimal environmental footprint, enhanced pest control, and provide high crop yield. It allows more accurate control of environmental conditions that offer possibilities for increasing production and improving the quality of crops. The rapid development of computers and controllers has enabled the opportunity to apply the controller in hydroponics. The microcontroller could be used to control these nutrient solution parameters by using relevant sensors. It monitors the conductivity and pH throughout 24 h during the whole cycle of production. Also, it helps in monitoring temperature, nutrient atomization, EC, and pH fluctuations and level of nutrient solution in the nutrient reservoir. However, although the comprehension of the multi-level interactions among the various mineral elements is considered crucial to understanding the different sensing and signaling pathways induced by a single or multiple shortage/s, the impact of these nutrients' interactions on crop performance is largely unknown [32].

Some hydroponic growers use more than the required amounts of nutrients for crop growth to minimize the chances of nutrient deficiency. But one of the most important factors for a successful hydroponics system is the use of the appropriate nutrient solution, and it is important to control the amount of nutrients to allow or deny plants the nutrient accumulation. While hydroponic systems are considered to represent a sustainable method for growing plants, the nutrient solution used in hydroponic systems is based on chemical fertilizers which are mined from scarce and non-renewable resources. Recently, there has been an increased interest in organic hydroponics, as the market for organic food continues to grow and some studies have reported the possibility of growing vegetables using an organic nutrient solution. For optimizing the utilization of organic waste for hydroponic plant growth, a solubilization step is required to break down organic matter and mobilize nutrients [96]. For example, the direct use of organic fertilizers in hydroponic systems may inhibit plant growth due to the high biological oxygen demand in the root zone caused by the presence of dissolved organic carbon compounds. Additionally, most of the nutrients in organic sources, such as waste material from the agricultural and aquacultural industry, are not in ionic forms and, hence, are not directly available for plants. Also, the last decade has seen increasing interest in using wastewater as a source of hydroponic nutrition. This aims at a dual benefit of optimizing water reuse as well as a practical end point for wastewater management. Untreated domestic rinse water obtained from washing machine second-cycle rinse can effectively be used for indoor hydroponic cultivation of plants without the need for any additional fertilizer. It also entails the benefits of significant savings in water use, sewage disposal, ecosystem protection, and the possibility to produce economically viable food crops [97]. Nowadays, hydroponic farming technology is extensively used in producing ornamental plants and flowers. Controllable application of fertilizers, the ability to change nutrients in different weather conditions and different plant growth stages, reduction of fertilizer leaching from the root zone, reduction of contamination, environmental protection, and enhancement of the quality and quantity of products are becoming some of the advantages of this technology.

Author details

Moaed Ali Al Meselmani
Grantham Centre for Sustainable Futures, The University of Sheffield, Sheffield, UK

*Address all correspondence to: m.almeselmani@sheffield.ac.uk

References

[1] Singh S, Singh BS. Hydroponics – A technique for the cultivation of vegetables and medicinal plants. In: Proceedings of 4th Global Conference on Horticulture for Food, Nutrition and Livelihood Options. Bhubaneshwar, Odisha, India; 2012. p. 220

[2] Sundara KMR, Nayagi DS, Jeevitha R, Veena K. Design and development of automatic robotic system for vertical hydroponic farming using IOT and big data analysis. Turkish Journal of Computer and Mathematics Education. 2021;**12**(11):1597-1607

[3] Levine CP, Mattson NS. Potassium-deficient nutrient solution affects the yield, morphology, and tissue mineral elements for hydroponic baby leaf spinach (*Spinacia oleracea* L.). Horticulturae. 2021;**7**:213. DOI: 10.3390/horticulturae7080213

[4] Yang T, Samarakoon U, Altland J, Ling P. Photosynthesis, biomass production, nutritional quality, and flavor-related phytochemical properties of hydroponic-grown arugula (*Eruca sativa* Mill.) 'standard' under different electrical conductivities of nutrient solution. Agronomy. 2021;**11**:1340. DOI: 10.3390/agronomy11071340

[5] Valentinuzzi F, Pii Y, Vigani G, Lehmann M, Cesco S, Mimmo T. Phosphorus and iron deficiencies induce a metabolic reprogramming and affect the exudation traits of the woody plant *Fragaria×ananassa*. Journal of Experimental Botany. 2015;**66**:6483-6495. DOI: 10.1093/jxb/erv364

[6] Nguyen VQ, Van HT, Le SH, Nguyen TH, Nguyen HT, Lan NT, et al. Production of hydroponic solution from human urine using adsorption–desorption method with coconut shell-derived activated carbon. Environmental Technology and Innovation. 2021;**23**: 101708

[7] Salisbury FB, Ross CW. Plant Physiology. 4th ed. Beverly: Wadsworth Publishing Company; 1991. p. 481

[8] Cooper A. 1. The system. 2. Operation of the system. In: Books G, editor. The ABC of NFT. Nutrient Film Technique. London; 1988. pp. 3-123

[9] Steiner AA. The universal nutrient solution. In: Proceedings of IWOSC 6th International Congress on Soilless Culture; Wageningen, The Netherlands; 1984. pp. 633-650.

[10] Windsor G, Schwarz M. Soilless Culture for Horticultural Crop Production. FAO, Plant Production and Protection. Roma, Italia; 1990. Unipub, ISBN: 0987650XXX.

[11] Hoagland DR, Arnon. The Water-Culture Method for Growing Plants Without Soil (Circular (California Agricultural Experiment Station), 347. ed.). Berkeley, CA: University of California, College of Agriculture, Agricultural Experiment Station; 1938

[12] Hewitt EJ. Sand and water culture methods used in the study of plant nutrition. In: Technical communication No. 22. East Malling, Maidstone, Kent, England: Commonwealth Bureau of Horticulture and Plantation Crops; 1996

[13] Hoagland DR, Snyder WC. Nutrition of strawberry plant under controlled conditions. (a) Effects of deficiencies of boron and certain other elements, (b) susceptibility to injury from sodium salts. Proceedings of the American Society for Horticultural Science. 1933;**30**:288-294

[14] Steiner AA. A universal method for preparing nutrient solutions of a certain desired composition. Plant and Soil. 1961;**15**(2):134-154. Available from: https://edepot.wur.nl/309364

[15] Bollard EG. A comparative study of the ability of organic nitrogenous compounds to serve as sole sources of nitrogen for the growth of plants. Plant and Soil. 1966;**25**:153-166

[16] Steiner AA. Soilless Culture, Proceedings of the IPI 1968 6th Colloquium of the International Potash Institute. Switzerland: International Potash Institute; 1968. pp. 324-341

[17] Salisbury FB, Ross CW. Plant Physiology. California: Wadsworth Publishing Company; 1992

[18] Taiz L, Zeiger E. Plant Physiology. Sunderland: Sinauer Associates; 1998

[19] Tahereh TA, Tabatabaei SJ, Torkashvand AM, Khalighi A, Talei D. Effects of silica nanoparticles and calcium chelate on the morphological, physiological and biochemical characteristics of gerbera (*Gerbera jamesonii* L.) under hydroponic condition. Journal of Plant Nutrition. 2021;**44**(7):1039-1053. DOI: 10.1080/01904167.2020.1867578

[20] Voogt W. Potassium management of vegetables under intensive growth conditions. In: Pasricha NS, Bansal SK, editors. Potassium for Sustainable Crop Production. Bern, Switzerland: International Potash Institute; 2002. pp. 347-362

[21] Modu F, Adam A, Aliyu F, Mabu A, Musa M. A Survey of smart hydroponic systems. advances in science. Technology and Engineering Systems Journal. 2020;**5**(1):233-248

[22] Sardare MD, Admane SV. A Review on plant without soil–Hydroponics. International Journal of Research in Engineering and Technology. 2013;**2**(3):299-304

[23] Sheng-Xiu L, Wang Z-H, Malhi SS, Li S-Q, Ya-Jun G, Xiao-Hong T. Nutrient and water management effects on crop production, and nutrient and water use efficiency in dryland areas of China. In: Sparks, editor. Advances in Agronomy. Vol. 102. Burlington: Academic Press; 2009. pp. 223-265

[24] Chrysargyris A, Petropoulos SA, Prvulovic D, Tzortzakis N. Performance of hydroponically cultivated geranium and common verbena under salinity and high electrical conductivity levels. Agronomy. 2021;**11**:1237. DOI: 10.3390/agronomy11061237

[25] Fageria VD. Nutrient interactions in crop plants. Journal of Plant Nutrition. 2021;**24**(8):1269-1290. DOI: 10.1081/PLN-100106981

[26] Marschner P. Rhizosphere biology. In: Marschner's Mineral Nutrition of Higher Plants. 3rd ed. Amsterdam: Academic Press; 2012. pp. 369-388

[27] Zhang F, Niu J, Zhang W, Chen X, Li C, Yuan L, et al. Potassium nutrition of crops under varied regimes of nitrogen supply. Plant and Soil. 2010;**335**:21-34. DOI: 10.1007/s11104-010-0323-4

[28] Valentinuzzi F, Pii Y, Mimmo T, Savini G, Curzel S, Cesco S. Fertilization strategies as a tool to modify the organoleptic properties of raspberry (*Rubus idaeus* L.) fruits. Scientia Horticulture. 2018;**240**:205-212. DOI: 10.1016/j. scienta.2018.06.024

[29] Silberbush M, Ben-Asher J. The effect of NaCl concentration on NO_3^-, K^+ and orthophosphate-P influx to peanut roots. Scientia Horticulturae. 1989;**39**:279-287. DOI: 10.1016/0304-4238(89)90121-0

[30] Schimansky CD. einfluß einiger versuchsparameter auf das fluxverhalten von 28mg bei gerstenkeimpflanzen in hydrokulturversuchen. Landwirtschaftliche Forschung. 1981;**34**:154-165

[31] Ranade-Malvi U. Interaction of micronutrients with major nutrients with special reference to potassium. Karnataka Journal of Agricultural Sciences. 2011;**24**(1):106-109

[32] Astolfi S, Celletti S, Vigani G, Mimmo T, Cesco S. Interaction between sulphur and iron in plants. Frontiers in Plant Science. 2021;**12**:670308. DOI: 10.3389/fpls.2021.670308

[33] Santamaria P. Nitrate in vegetables: Toxicity, content, intake and EC regulation. Journal of the Science of Food and Agriculture. 2006;**86**:10-17. DOI: 10.1002/jsfa.2351

[34] Savvas D. Hydroponics: A modern technology supporting the application of integrated crop management in greenhouse. Food, Agriculture and Environment. 2003;**1**:80-86

[35] Sambo P, Nicoletto C, Giro A, PII Y, Valentinuzzi F, Mimmo T, et al. Hydroponic solutions for soilless production systems: Issues and opportunities in a smart agriculture perspective. Frontiers in Plant Science. 2019;**10**:923

[36] Nacry P, Bouguyon E, Gojon A. Nitrogen acquisition by roots: Physiological and developmental mechanisms ensuring plant adaptation to a fluctuating resource. Plant and Soil. 2013;**270**:1-29. DOI: 10.1007/s11104-013-1645-9

[37] Ferrón-Carrillo F, da Cunha-Chiamolera TPL, Urrestarazu M. Effect of ammonium nitrogen on pepper grown under soilless culture. Journal of Plant Nutrition. 2021;**45**:113-122. DOI: 10.1080/01904167.2021.1943438

[38] Kronzucker HJ, Glass ADM, Siddiqi MY. Inhibition of nitrate uptake by ammonium in barley. Analysis of component fluxes. Plant Physiology. 1999;**120**:283-292. DOI: 10.1104/pp.120.1.283

[39] Forde BG, Clarkson DT. Nitrate and ammonium nutrition of plants: Physiological and molecular perspectives. Advances in Botanical Research. 1999;**30**:1-90. DOI: 10.1016/S0065-2296(08)60226-8

[40] Bloom AJ, Jackson LE, Smart DR. Root growth as a function of ammonium and nitrate in the root zone. Plant, Cell & Environment. 1993;**16**:199-206. DOI: 10.1111/j.1365-3040.1993.tb00861.x

[41] Ahmadi F, Samadi A, Sepehr E, Rahimi A, Shabala S. Perlite particle size and NO3 – /NH4+ ratio affect growth and chemical composition of purple coneflower (*Echinacea purpurea* L.) in hydroponics. Industrial Crops and Products. 2021;**162**:113285

[42] Lee JY, Rahman A, Azam H, Kim HS, Kwon MJ. Characterizing nutrient uptake kinetics for efficient crop production during *Solanum lycopersicum* var. cerasiforme Alef. growth in a closed indoor hydroponic system. PLoS One. 2017;**12**:e0177041. DOI: 10.1371/journal.pone.0177041

[43] Packter A. The precipitation of calcium sulphate dihydrate from aqueous solution: Induction periods, crystal numbers and final size. Journal of Crystal Growth. 1974;**21**:191-194. DOI: 10.1016/0022-0248(74)90004-9

[44] McBride MB. Environmental Chemistry of Soils. New York, USA: Oxford University Press; 1994

[45] Lucena JJ. Fe chelates for remediation of Fe chlorosis in strategy I plants. Journal of Plant Nutrition. 2003;**26**:1969-1984. DOI: 10.1081/PLN-120024257

[46] Truog E. Soil reaction influence on availability of plant nutrients. Soil Science Society of America Proceedings. 1964;**11**:305-308

[47] Mayavan RRS, Jeganath R, Chamundeeswari V. Automated hydroponic system for deep water culture to grow tomato using atmega328. In: Proceedings of Technoarete International Conference. Chennai, India; 2017

[48] Sharma N, Acharya S, Kumar K, Singh N, Chaurasia OP. Hydroponics as an advanced technique for vegetable production: An overview. Journal of Soil and Water Conservation. 2018;**17**(4):364-371. DOI: 10.5958/2455-7145.2018.00056.5

[49] Gillespie DP, Papio G, Kubota C. High nutrient concentrations of hydroponic solution can improve growth and nutrient uptake of spinach (*Spinacia oleracea* l.) grown in acidic nutrient solution. HortScience. 2021;**56**(6):687-694. DOI: 10.21273/HORTSCI15777-21

[50] Islam AKMS, Edwards DG, Asher CJ. PH optima for crop growth: Results of a flowing solution culture experiment with six species. Plant and Soil. 1980;**54**:339-357

[51] Frick J, Mitchell CA. Stabilization of pH in solid matrix hydroponic systems. HortScience. 1993;**28**:981-984. DOI: 10.21273/hortsci.28.10.981

[52] Marschner H. Mineral Nutrition of Higher Plants. Cambridge, Massachusetts: Academic Press; 1995

[53] Gillespie DP, Kubota C, Miller SA. Effects of low pH of hydroponic nutrient solution on plant growth, nutrient uptake, and root rot disease incidence of basil (*Ocimum basilicum* L.). HortScience. 2020;**55**:1251-1258. DOI: 10.21273/HORTSCI14986-20

[54] Voogt W. Effects of the pH on rockwool grown carnation (*Dianthus caryophyllus*). Acta Horticulturae. 1995;**401**:327-336

[55] Takeno N. Atlas of Eh-pH Diagrams. Intercomparison of Thermodynamic Databases. Tokyo: National Institute of Advanced Industrial Science and Technology; 2005. p. 285

[56] Cerozi da Silva B, Fitzsimmons K. The effect of pH on phosphorus availability and speciation in an aquaponics nutrient solution. Bioresource Technology. 2016;**219**:778-781

[57] Resh HM. Hydroponic food production. 6th ed. Mahwah, NJ: New Concept Press Publishing Co.; 2004

[58] Bugbee B. Nutrient Management in Recirculating Hydroponic Culture. 2003. Available from: http://www.usu.edu/cpl/researchhydroponics3.htm Accessed: November 22, 2010

[59] White PJ. Ion uptake mechanisms of individual cells and roots: Short distance transport. In: Marschner P, editor. Marschner's Mineral Nutrition of Higher Plants. London: Academic Press; 2012. pp. 7-47

[60] Asher CJ, Edwards DG. In: Pirson A, Zimmermann MH, editors. Modern Solution Culture Techniques: In Inorganic Plant Nutrition. Berlin, Heidelberg: Springer; 1983. pp. 94-119

[61] Alexopoulos AA, Marando E, Assimakopoulou A, Vidalis N, Petropoulos SA, Karapanos IC. Effect of Nutrient Solution pH on the Growth,

Yield and Quality of *Taraxacum officinale* and *Reichardia picroides* in a Floating Hydroponic System. Agronomy. 2021;**11**:1118. DOI: 10.3390/agronomy11061118

[62] Cole DL, Kobza SJ, Fahning SR, Stapley SH, Bonsrah DKA, Buck RL, et al. Soybean nutrition in a novel single-nutrient source hydroponic solution. Agronomy. 2021;**11**:523. DOI: 10.3390/agronomy 11030523

[63] Wang L, Chen X, Guo W, Li Y, Yan H, Xue X. Yield and nutritional quality of water spinach (*Ipomoea Aquatica*) as influenced by hydroponic nutrient solutions with different pH adjustments. International Journal of Agriculture and Biology. 2017;**19**:635-642

[64] Dickson RW, Fisher PR. Quantifying the acidic and basic effects of vegetable and herb species in peat-based substrate and hydroponics. Hort Science. 2019;**54**:1093-1100. DOI: 10.21273/HORTSCI13959-19

[65] Grattan SR, Grieve CM. Salinity-mineral nutrient relations in horticultural crops. Scientia Horticulturae. 1999;**78**:127-157

[66] Sonneveld C, Voogt W. Plant Nutrition of Greenhouse Crops. Dordrecht: Springer; 2009. p. 403

[67] Lenni HS, Seminar KB, RPA S. Development of a control system for lettuce cultivation in floating raft hydroponics. IOP Conference Series: Earth and Environmental Science. 2020;**542**(1):012067

[68] Zhang P, Senge P, Dai Y. Effects of salinity stress on growth, yield, fruit quality and water use efficiency of tomato under hydroponics system. Reviews in Agricultural Science. 2016;**4**:46-55

[69] Samarakoon UC, Weerasinghe PA, Weerakkody WAP. Effect of electrical conductivity (ec) of the nutrient solution on nutrient uptake, growth, and yield of leaf lettuce (*Lactuca sativa* L.) in stationary culture. Tropical Agriculture Research. 2006;**18**:13-21

[70] Lykas CN, Katsoullas P, Giaglaras P, Kittas C. Electrical conductivity and pH prediction in Recirculated nutrient solution of greenhouse soilless rose crop. Journal of plant nutrition. 2006;**29**:1585-1599

[71] Muthir SA, Salim AA, Yaseen AA, Saleem KN. Influence of nutrient solution temperature on its oxygen level and growth, yield and quality of hydroponic cucumber. Journal of Agricultural Science. 2019;**11**(3):75-92

[72] Nxawe S, Ndakidemi PA, Laubscher CP. Possible effects of regulating hydroponic water temperature on plant growth, accumulation of nutrients and other metabolites. African Journal of Biotechnology. 2010;**9**(54):9128-9134

[73] Aðalsteinsson S, Jensén P. Influence of temperature on root development and phosphate influx in winter wheat grown at different P levels. Physiologia Plantarum. 2006;**80**:69-74

[74] Xu Q, Huang B. Seasonal changes in root metabolic activity and nitrogen uptake for two cultivars of creeping bentgrass. Horticultural Science. 2006;**41**:822-826

[75] Yamori W, Noguchi K, Hanba TK, Terashima I. Effects of internal conductance on the temperature dependence of the photosynthetic rate in Spinach leaves from contrasting growth temperatures. Plant & Cell Physiology. 2006;**47**:1069-1080

[76] Calatayud A, Gorbe E, Roca D, Martínez PF. Effect of two nutrient solution temperatures on nitrate uptake, nitrate reductase activity, NH_4^+ concentration and chlorophyll a fluorescence in rose plants. Environmental and Experimental Botany. 2008;**64**(1):65-74

[77] Daskalaki A, Burrage SW. Solution temperature and the uptake of water and nutrients by cucumber (*Cucumis sativus* L.). Acta Horticulturae. 1998;**458**:317-322

[78] Reyes DM, Stolzy LH, Labanauskas CK. Temperature and oxygen effects in soil on nutrient uptake in Jojoba seedlings. American Society of Agronomy. 1977;**69**:647-650

[79] Olivier OJ. All-year-round vegetable production without protection and future prospects in South Africa. Acta Horticulturae. 1974;**42**:353-362

[80] Mills PJW, Smith IE, Marais G. A greenhouse design for a cool subtropical climate with mild winters based on microclimatic measurements of protected environments. Acta Horticulturae. 1990;**281**:83-94

[81] Fazlil Ilahi WF, Ahmad D, Husain MC. Effects of root zone cooling on butterhead lettuce grown in tropical conditions in a coir-perlite mixture. Horticulture, Environment and Biotechnology. 2017;**58**:1-4. DOI: 10.1007/s13580-017-0123-3

[82] Kozai T. Closed systems for high quality transplants using minimum resources. Plant Tissue Culture Engineering. 2006;**6**:275-312

[83] Sethi VP, Sharma SK. Greenhouse heating and cooling using aquifer water. Energy. 2007;**32**:1414-1421

[84] Pezeshki SR, Pardue JH, Delaune RD. The influence of soil oxygen deficiency on alcohol dehydrogenase activity, root porosity, ethylene production and photosynthesis in Spartina patens. Environmental and Experimental Botany. 1993;**33**(4):565-573

[85] Libia ITT, Gómez-Merino FC. In: Asao T, editor. Nutrient Solutions for Hydroponic Systems, Hydroponics—A Standard Methodology for Plant Biological Researches. Rijeka: InTech; 2012. Available from: http://www.intechopen.com/books/hydroponics-a-standardmethodology-for-plant-biological-researches/nutrient-solutions-for-hydroponic-systems

[86] Jong WL, Beom SL, Jong GK, Jong HB, Yang GK, Shela G, et al. Effect of root zone aeration on the growth and bioactivity of cucumber plants cultured in perlite substrate. Biologia. 2014;**69**(5):610-617. DOI: 10.2478/s11756-014-0360-1

[87] Roosta HR, Bagheri MH, Hamidpour M, Roozban MR. Interactive effects of nitrogen form and oxygen concentration on growth and nutritional status of eggplant in hydroponics. Journal of Agricultural Science and Technology. 2016;**18**:731-739

[88] Lemon ER, Wiegand CL. Soil aeration and plant root relations II root respiration. Agronomy Journal. 1962;**54**:171-175

[89] Cherif M, Tirilly Y, Belanger RR. Effect of oxygen concentration on plant growth, lipid peroxidation, and receptivity of tomato roots to Pythium F under hydroponic conditions. European Journal of Plant Pathology. 1997;**103**(3):255-264

[90] Morard P, Silvester J. Plant injury due to oxygen deficiency in the root environment of soilless culture: A Review. Plant and Soil. 1996;**184**(2):243-254

[91] RDSC C, Bastos RG, Souza CF. Influence of the use of wastewater on nutrient absorption and production of lettuce grown in a hydroponic system. Agricultural Water Management. 2018;**203**:311-321. DOI: 10.1016/j.agwat.2018.03.028

[92] Bar-Yosef BT. In: Raviv M, Lieth HJ, editors. Fertigation Management and Crops Response to Solution Recycling in Semi-Closed Greenhouses: In Soilless Culture: Theory and Practice. Amsterdam, The Netherlands: Elsevier; 2008. pp. 341-424

[93] Carmassi G, Incrocci L, Maggini R, Malorgio F, Tognoni F, Pardossi A. Modeling salinity build-up in recirculating nutrient solution culture. Journal of Plant Nutrition. 2005;**28**:431-445. DOI: 10.1081/PLN-200049163

[94] Parida AK, Das AB. Salt tolerance and salinity effects on plants: A review. Ecotoxicology and Environmental Safety. 2005;**60**:324-349. DOI: 10.1016/j.ecoenv.2004.06.010

[95] Baiyin B, Tagawa K, Yamada M, Wang X, Yamada S, Yamamoto S, et al. Effect of the flow rate on plant growth and flow visualization of nutrient solution in hydroponics. Horticulturae. 2021;**7**:225. DOI: 10.3390/horticulturae7080225

[96] Ezziddine M, Liltved H, Seljåsen R. Hydroponic lettuce cultivation using organic nutrient solution from aerobic digested aquacultural sludge. Agronomy. 2021;**11**:1484. DOI: 10.3390/agronomy11081484

[97] Sundar P, Jyothi K, Sundar C. Indoor hydroponics: A potential solution to reuse domestic rinse water. Biosciences Biotechnology Research Asia. 2021;**18**(2):373-383

Chapter 3

A Potential Alternative for Agar in In Vitro Culture Media Based on Hydrocolloids Present in Nopal: General Structure and Mechanical Properties

Arantza Elena Sánchez-Gutiérrez,
Genaro Martín Soto-Zarazúa, Manuel Toledano-Ayala
and Juan Fernando García-Trejo

Abstract

In Vitro culture is a technique commonly used for plant research. Nevertheless, it is more expensive than traditional methods of production, due to the use of the culture medium gelling agent called agar. Recent studies have been searching for alternative substances in raw materials with the same characteristics but which can be extracted easier than agar. The dietary fiber of the nopal cactus (Opuntia) is a rich source of hydrocolloids (pectin and mucilage). These hydrocolloids have the ability to gel in combination with the indicated solution. In this chapter, we will focus on the study of the hydrocolloids from nopal cactus to replace agar partially and/or totally as a gelling agent using in vitro culture media benefiting from the molecular structure and mechanical properties of the compounds.

Keywords: In vitro culture media, gelling agents, mechanical properties, molecular structure, nopal cactus hydrocolloids

1. Introduction

Due to the growing global crisis caused by climate change and greenhouse gases, which are significantly affecting food availability and price volatility in the agricultural sector, agricultural systems have been forced to implement new adaptive measures with the aim of developing alternatives that improve and accelerate traditional methods of food production through biotechnology [1, 2].

Plant biotechnology had an important development in the last decades, moving from soil cultivation to practices where ecological factors are involved with sustainable objectives such as water conservation, decrease of soil erosion, and higher yields and better-quality crops. As a result of this movement, the technique of in vitro culture arose, in which plants are propagated inside containers under laboratory conditions [3] and has open the possibility of enhancing two conditions in agricultural crops, firstly, the asepsis, which is the absence of germs, and secondly, the control of factors that affect plants growth.

However, although it is a technique that has been widely used in agricultural development and plant research, the in vitro culture technique is more expensive than traditional methods of plant production, due to the components used for the preparation of the culture medium, commonly the gelling agent called agar [4–6].

The culture media must meet the nutritional requirements necessary for the development of the plant and these depend on the culture used, which is in some cases solidified to provide support to the explant through the gelling agent agar. Agar is a mixture of polysaccharides extracted from the walls of red algae (Gelidium, Gracilaria, Pterocladia, Gelidiella) and is the most widely used gelling agent in culture media due to its physicochemical properties such as porosity, thermo-reversibility, stability, gel strength, texture, elasticity, and transparency. Its use increases the total cost of the medium by 70%, due to agar overexploitation and high demand [7].

As a result of this, recent studies have made it possible to know with greater precision the availability of alternative substances to agar in naturally occurring raw materials that are easier to extract and have the same characteristics as agar. The study of natural polymers of vegetable origin has emerged as a sustainable alternative due to their properties. Starches and gums have been investigated due to their high availability in local markets and low cost [8–10], concluding that they can potentially replace agar partially and/or totally as gelling agents due to their physicochemical composition, and can reduce costs of this technology due to their greater efficiency, ease of extraction and acquisition [11].

The nopal cactus (Opuntia ficus-indica), a native plant distributed throughout Latin America, is a natural polymers supplier that has been studied in numerous investigations for its properties, since phenolic compounds had been found in the composition, which can be used in various industries, including pharmaceutical, construction, agrotechnology, bioenergy, and biotechnology industries [12]. One of the most important components of nopal cactus is dietary fiber, which is a rich source of hydrocolloids (pectin and mucilage), named as such because of their great capacity to capture and retain water.

Pectin is mainly composed of galacturonic acid, and mucilage is composed of arabinose, galactose, xylose, rhamnose, and galacturonic acid (classified as an acid mucilage) [13]. These are hydrocolloids that have the ability to gel and form gels in combination with the indicated solution. In addition, it is well known that one of the main functions of mucilages is to promote seed germination since when they come into contact with water, they increase their volume, forming a moist layer around the seed, which facilitates germination and protects the plant from external damage [14, 15].

Recently, nopal cactus hydrocolloids have been used as a thickening, stabilizing, encapsulating, and moisturizing agents in different research projects; however, in this chapter we will focus on the study of the nopal cactus hydrocolloids for their potential to substitute (partially or fully) the gelling agent agar for in vitro culture media applications, benefiting from the molecular structure and the mechanical properties of those compounds.

First, we will describe in vitro culture and the important features for the development of this technique, followed by the gelling agents and hydrocolloids mostly used to improve the technique and the recently studied substitutes, including their characteristics. Finally, we will discuss the potential application of nopal cactus

DOI: http://dx.doi.org/10.5772/intechopen.101745

hydrocolloids as gelling agents in in vitro culture and describe their molecular structure and mechanical properties.

2. The history of in vitro culture

The term "in vitro tissue culture" means growing explants in a glass bottle in an artificial environment in which asepsis, growth, and development must be controlled. Plant culture has undergone a significant evolution in the last 100 years. The knowledge and research related to cell theory has reached a significant development, which has generated a disruptive innovation in traditional culture.

The history of in vitro culture began in 1887 with Schleiden and Schwan, who explained that a cell could subsist by itself if the external conditions were favorable for its growth (cell theory). A decade later, in 1902 Herbertland, pioneer of in vitro culture and called the father of the technique, said that plants were capable of reproducing their growth from isolated cells. Herbertland proposed that it was possible to grow free vegetative cells and pollen tubules together by adding nutrient solutions supplemented with extracts of vegetative apices or with fluids from embryo sacs, and although he could not prove it due to the simplicity of his culture media, it was a breakthrough that gave the guideline to start investigating the technique [16].

In addition, Herberlat was the first to mention cell totipotency, which is defined as the ability of a cell to generate an individual completely identical to the mother cell from a single cell, which contains the same genetic material of the plant to which it belongs, therefore, it has the potential to generate a completely new plant [17]. By 1924, Blumenthal and Meyer demonstrated callus formation using carrot slices and lactic acid.

In the 1930's White, Gautheret and Nobécourt decisively demonstrated the possibility of cultivating plant cells in vitro, discovering two important characteristics, the identification of auxins as regulators of plant growth and the importance of B complex vitamins in plant growth [18]. By 1940 Blakeslee, Conklin, and Van Overbeek studied a semisynthetic liquid medium with coconut milk, which had a good proliferation response, which promoted biochemical research.

In 1962 Murashige and Skoog made a medium for tobacco tissues that contained all the nutrients necessary for the growth and proliferation of cays, and a high concentration of salts that benefited the growth of somatic embryos [19]. Subsequently, research was carried on somatic embryogenesis, explant types, the culture of microspores and meristems, and obtaining hybrids.

Culture media had been an important object of study, in 1984, Wetherell proposed that high levels of sucrose up to 12%, nitrate, and iron are essential for somatic embryos to develop to maturity [20]. However, nowadays most culture media consist of five groups of ingredients: organic nutrients, carbon source, vitamins, and growth regulators such as phytohormones and their inhibitors (auxins, gibberellins, cytokinins, abscisic acid, ethylene, brassinosteroids, polyamines, jasmonic acid, and salicylic acid).

Auxins promote cell elongation and cytokinins stimulate cell division [20]. These media are mostly used to promote organogenesis and their composition is determinant for growth. The Murashige & Skoog (MS) medium designed for tobacoo cells is the most used due to its success for agar culture [21] although there are other media such as Schen and Hildebrandt, Heller, MS, and Eriksson, among others, which contain certain macro and micronutrient content that are also used for plant cell culture.

Another factor that influences the growth of tissues in solid culture media is their consistency, which is the result of the selection of the gelling agent [22]. There is a great variety of gelling agents, the most used are agar, agarose, gellan gum, or calcium alginate [23]. Agar is commonly used in proportions of 0.6% to 1.0%, due to its composition, purity, and properties, which do not interfere with the growth of the culture [24].

3. Commercial gelling agents

Gelling agents such as agar-agar, phytagel®, gelrite® or gellam gum, natugel®, agarose, alginate and isabgol, and gums that have been combined to be substitutes for agar, such as guar gum, cassia, xanthan gum, and katyra gum, which function as thickening, gelling, foaming and stabilizing agents, are available commercially (**Table 1**).

Gelling agent	Chemical Composition	Category	Application	Cost
Phytagel / Gelrite/ Gellam gum	Fermentation product of various Pseudomonas species	Heteropolysaccharides secreted by bacteria of the genus Sphinogomonaso.	Plant tissue culture medium. Agar substitute	250 g 90.79 USD
Natugel	Seaweed Kappaphycus alvarezii	Increased viscosity and consistency	Gel-forming agents, use in various research areas	250 g 225.81 USD
Agarose	Consisting of repeating units of a molecule called agarobiose.	Polysaccharide from algae of the genera Gellidium and Gracillaria.	Gels that allow the separation of DNA molecules by electrophoresis, cell culture, and microbiology.	100 g 245.78 USD
Alginate	Organic polymers derived from alginic acid.	Anionic polysaccharide from brown algae	Gelling and spherification agent	454 g 73.69 USD
Isubgol	It comes from the mucilage of Plantago ovata seeds.	Plantago Ovata seed hulls, rich in fiber with high mucilage content.	Gelling agent, diuretic, weight loss.	220gr 23.59 USD
Guar	Galactose (bound by α1–6) and mannose (bound by β1–4).	Seed flours Cyamopsis tetragonolobus	Stabilizer, thickener, and emulsifier. Gelling agent together with isubgol.	227 g 9.85 USD
Cassia	Galactose (bound by α1–6) and mannose (bound by β1–4).	Sennao btsifolia seed	Thickener and gelling agent in combination with agar.	150 g 10.07 USD
Xanthan	D-glucopyranosyl chain linked β1–4 bond, branching of D-mannopyranosyl trisaccharides and D-glucopyranosyluronic acid.	Fermentation Exudate of the bacterium Xanthomonas campestris B-1459	Stabilizer, thickener, emulsifier, emulsifying agent, stabilizing agent, thickening agent, emulsifying agent to provide shape and foaming and gelling agent.	250 g 12.68 USD

Gelling agent	Chemical Composition	Category	Application	Cost
Agar	Agarose (galactoses) and agaropectin (anhydrogalactose partially esterified with sulfuric acid)	Gelidium cartilagineum, Gracilaria confervoides and Peteroclaia capillacea algae extract.	Stabilizer, thickener, emulsifier and gelling agent.	75 g 326.24 USD
Katira Rubber	D-galactose, D-galacturonic acid and L-rhamnose in molar ratio 2:1:3	Plant extract Exudate of Cochlospermun religiosum	Emulsifier or thickener and gelling agent.	100 g 680.29 USD

Table 1.
Commercial gelling agents. Own elaboration.

4. Agar

The development of solid culture media was fundamental not only for bacteriology but also for biotechnology. Robert Koch and his assistant Walter Hesse are known as the inventors of solid media, in 1882 they replaced animal gelatin with agar gelatin, however, this technique already had precursors. Some authors mentioned that agar was discovered in Japan in 1658 by Tarazaemon Minoya, agar was used at the time as a gelling agent for food [25]. Agar is extracted from the cell membrane wall of red algae in a fibrous crystallized form and is mainly composed of sulfated galactan [26].

Around the world, there are different genera of algae such as Gelidium, Gracilaria, Gelidiella, Pterocladia, Gracilariopsis, Ahnfeltia, and their respective species. The characteristics of each of these provide the extracted agar with different properties. However, the best quality agar is derived from Gelidium, algae that are only found in the wild. It is important to mention that the composition of the agar is not affected by seasonality, but the yield is [27].

Agar is classified as a non-branched, high molecular weight, reserve natural polysaccharide with mineral content including Copper (Cu), Zinc (Zn), Manganese (Mn), Calcium (Ca), and Magnesium (Mg). Agar is composed of agaropectin and agarose, the latter is the main constituent (70%). Both components have the same basic structure, formed by alternating units of D-galactose and 3,6-anhydro-L-galactopyranose linked by α-(1–3) and β-(1–4) bonds [28].

Agarose has the following formula $[C_{12}H_{14}O_5(OH)_4]$, and is classified as a neutral polysaccharide. The α and β galactoses present in their molecular structure are responsible for the mechanical property of gelation which is linked to the formation of gels, caused by the interaction of helicoids. Agarose and agaropectin are differentiated by the presence of sulfate and pyruvate residues. Agaropectin is classified as a charged polymer, where D-galacturonic acid and pyruvic acid provide it with viscosity property [26, 29]. The aforementioned composition is similar to that of the natural polymer starch [30].

4.1 General structure and properties of agar

The FTIR spectrum of agar (Gelidium) has vibrations in the absorption bands at 3291–3390 cm^{-1} associated with O-H stretching, followed by vibration at 2932–2922 cm^{-1} for CH_2 stretching, and the absence of a band at 2845 cm^{-1} corresponding to the $O\text{-}CH_3$ groups indicates a degree of low methoxylation. A vibration in the

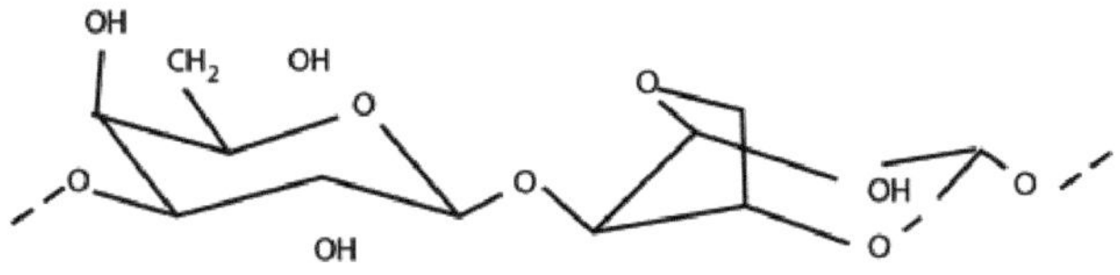

Figure 1.
Chemical structure of agar [33].

1642 cm^{-1} band is associated with the stretching vibration of the conjugated peptide bond formed by amine (NH) and acetone (CO) groups. Vibrations for CH_2 groups are found at 1413–1370 cm^{-1}, and a vibration that is identified for bridging a sulfated ester at 1179 cm^{-1}. In addition, the most represented vibration band in the agars is found at 930 cm^{-1} and attributed to the bridging of 3,6-anhydro-galactose [30–32]. A schematic of the agar structure is shown in **Figure 1**.

Agar is a hydrophilic hydrocolloid that has different properties, including the ability to form colorless, thermo-reversible gels that do not lose their original characteristics with changes in temperature; the gel forms at 30°C and dissolves at temperatures between 75°C and 90°C. The gels are not digested by plant enzymes. Important properties of these gels, such as yield and gel strength, may vary depending on the genus and location of the algae (**Table 2**).

For the extraction of agar there are different methodologies that can alter the above-mentioned properties (yield and gel strength). The most used methodologies are direct extraction in the water bath, the Freeze–Thaw method, syneresis method, and the alkaline treatment. Recent extraction processes are photobleaching, microwave-assisted extraction, and enzymatic method assisted by hydrogen peroxide [40, 41].

On the other hand, agar has important uses in different industries in addition to the elaboration of solid culture media for plant growth and micropropagation. It was the first hydrocolloid used as a gelling and stabilizing agent in the food industry. Agar is also used in the dental industry for the manufacture of the mold, in the medical industry for pharmaceutical formulations, in other industries for biodegradable films as packaging film; and its use in the cosmetic industry is well known [32].

The industry of agar production continues growing, Asia and the Pacific dominate the manufacture of this product, Indonesia and China play an important role, but Chile has prospered in that sense. The main suppliers of Gelidium are Spain, Portugal,

Species	Performance (dry weight) %	Gel strength ($g\ cm^{-2}$)	Author
Gracilaria verrucosa	4.3	225.8	Montilla-Escudero et al. [34]
Gracilaria cornea	36.6–46.1	251	Freile-Pelegrín et al. [35]
Gelidium coulteri	24.0–39.0	263–288 g	Macler & West [36]
Gelidium robustum	45.0–37.0	268–288	Hurtado et al. [37]
Gelidium sesquipedale	40.0–45.0	1000	Mouradi-Givernaud et al [38]
Gracilariopsis tenuifrons	39.57	1231	Zecchinel et al. [39]

Table 2.
Yield and strength properties of gel in different species of algae.

Morocco, Japan, Republic of Korea, China, Chile, and South Africa [42]. Annually, 55,000 tons of seaweed are extracted, which produce 7500 tons of agar with a value of US$ 132 million [43]. The growth of this industry has caused ecological problems that have hindered agar temporalities and supply; therefore, scientists have been inclined to find total or partial substitutes for agar for different industries. The following section explores some commercial gelling agents and natural agar substitutes with successful results.

4.2 Agar substitutes for gelling agents in culture media

Currently, research is seeking to reduce the ecological effects of agar extraction, while reducing the costs of this biotechnology [6]. Therefore, in 1986, the carrageenan was used as a substitute for the gelling agent agar in plant tissue cultures; according to the results obtained in the research, the tissues grew better on this gel than when the medium was solidified with agar [44].

In Cuba, the Center for Research and Development of Medicines (CIDEM) investigated the use of Aloe vera (Aloe barbadensis miller) and sago flour (Maranta arundinacea) as solid support in culture media for medicinal plants. The research demonstrated that partial or total substitution of agar by A. vera gel or sago flour is possible. In addition, this culture medium has been used for the in vitro propagation of agraz (Vaccinium meridionale Swartz) [45, 46].

Starches from cassava (Manihot esculenta), corn (Zea mays), and rice (Oryza sativa) have been investigated for their high availability in local markets and low cost. The use of starches in the partial replacement of Phytagel® in the modified MS medium for sweet potato (Ipomoea batatas) and cassava (M. esculenta) crops was investigated in Honduras at the Plant Tissue Culture Laboratory. As a result, it can replace up to 72% of the Phytagel® dose in sweet potato (I. batatas) and cassava (M. esculenta) crops.

Isabgol, which is the seed of Plantago psyllium, a herbaceous species from Spain and Morocco, used commercially for the production of mucilage for dietary fiber, in conjunction with commercial sugar was used as an alternative agar in in vitro culture media for plantain. The results showed that not only can isabgol be a solidifying agent in culture media, but also a preservation medium for germplasm banks [5].

In 2012, a study conducted in Ethiopia investigated the efficiency of ensete (bulla) starch as a gelling agent, significantly improving the number of sprouts and saving about 72.5% in costs [47]. In another study, the partial substitution of agar by the starch of the Diacol Capiro variety in the micropropagation of lulo Solanum quitoense Lam. was carried out, with a positive result [48].

Future perspectives in the development of this biotechnology are directed towards the partial or total substitution of agar as a gelling agent by solidifying agents that can be used more efficiently, easily extracted, locally acquired, and not temporary. In addition to improving plant production in reduced spaces and the resources derived to face future challenges with micropropagation without the risk of crop contamination.

5. Nopal hydrocolloids: potential substitutes of agar

Opuntia, better known as Nopal cactus is native to the American continent and belongs to the subfamily Opuntioideae (Cactaceae) that consists of 181 known species that are present throughout the American territory in the wild and are characterized by their easy reproduction and their ability to adapt to different climatic conditions [49]. Nopal cactus is produced worldwide and has been used by different industries in research related to natural medicine and human body benefits because it has essential nutrients for human beings such as dietary fiber, vitamin C and A, calcium, phosphorus, potassium, magnesium, chlorophyll, and antioxidants. The content of these nutrients is related to variables such as the age of the plant, the place where it grows, and the climate to which it has been exposed [50, 51].

However, it has been determined that the amino acid content of nopal cactus consists of aspartic acid, glutamic acid, serine, glycine, histidine, arginine, threonine, alanine, proline, tyrosine, valine, methionine, isoleucine, leucine, phenylalanine, lysine, and tryptophan [52]. Within its mineral composition, there is manganese, iron, zinc, magnesium, and sodium.

Dietary fiber is the major component of the nopal cactus, which can range from 11.0 g to 23.33 g depending on the age of the stalk and has been related to health benefits in different research studies due to its high content of bioactive compounds such as phenols and carotenoids [53]. Dietary fiber is composed of insoluble fiber and soluble fiber.

Soluble dietary fiber is formed by hydrocolloids (pectin and mucilage), which are named like that for their great capacity to capture and retain water [54]. These hydrocolloids are classified as natural polymers that have recently been studied for their importance and technological advantages. They are composed of arabinose, galactose, rhamnose, xylose, and galacturonic acid residues [55]. The mucilages of the nopal cactus have a similar composition to the exudates of Sterculia trees (Sterculia and Khaya gum), which are used as stabilizers [56].

However, due to the molecular composition and characteristics of the hydrocolloids present in the nopal cactus, this work revises definitions and characteristics of mucilage and pectin from nopal cactus stalks, the extraction methods, and mechanical properties.

5.1 Structure and general properties of mucilage

Mucilage is a complex neutral polymeric substance that is part of the carbohydrates and has a yield of 1.0% to 1.2% in fresh weight and about 17.9% in dry weight [57]. It is present in the Golgi apparatus and functions as a chelator capable of binding calcium and controlling the amount of free soluble calcium [58]. Mucilage has a branched chemical structure with approximately 55 sugars where L-arabinose, D-galactose, L-rhamnose, D-xylose, and D-galacturonic acid are present in different proportions [59–61].

Sáenz et al., [62] mention that McGarvie and Parolis (1981) presented the first suggested structure for O. ficus-indica mucilage, where they describe the molecule as a linear repeating central chain of α-D-linked (1–4) and β-L-linked (1–2) rhamnose with side chains of (1–6)-β-D-galacturonic acid linked to O-4 rhamnose residues. The galactose residues substituents at the O-3 positions, or double substituents at O-3 and O-4 (**Figure 2**) [63, 64].

Sepúlveda [57] mentions that the structure of mucilage is proposed as two distinct water-soluble fractions, where one is identified as pectin with gelling properties with $Ca^{2+}$and the other is a mucilage without gelling properties that swells when dissolved in water and shows characteristics of high viscosity [61, 65].

The property of viscosity, a physical characteristic of fluids, has a relationship to ionic strength, pH, and slightly to temperature in the Opuntia spp. As pH increases from acidic to alkaline conditions, viscosity increases. In addition, viscosity decreases

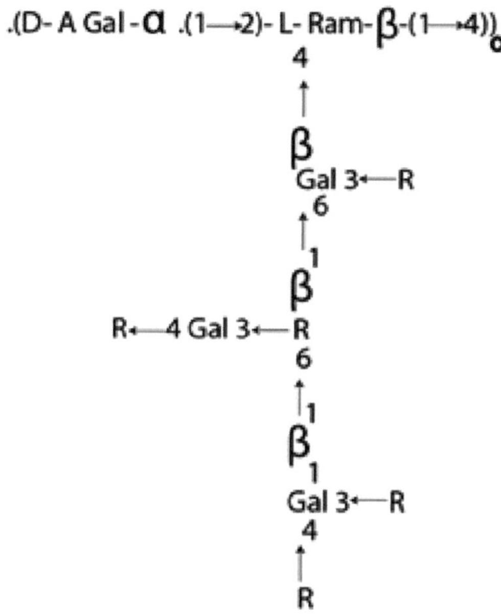

Figure 2.
Partial structure proposed for the mucilage of Opuntia ficus-indica R = arabinose or xylose [63].

as temperature increases just as it does in Xanthan gum. It is also mentioned that Opuntia mucilage has high elastic properties, the higher the concentration of mucilage, the lower the normal stresses [60]. The concentration of mucilage is important in the characterization of certain properties.

The extraction of mucilage can be carried out by different methods, and its yield depends on this. First, there is the extraction by water bath and the use of ethanol (95%) or isopropyl alcohol (95%), with a yield of 1.58% fresh weight [57]. The microwave-assisted extraction, the mechanical pressing system, and its lyophilization are other extraction methods.

The FTIR analysis of the mucilage of O. ficus indica and O. robusta performed in this study, compared with that performed in other species such as O. jonostle, O. streptacantha, O. tomentosa, O. atropes, and O. hyptiacantha shows that the functional groups present in all cactus mucilages are: absorption bands at 3500–3200 cm^{-1} that represent the carboxylic acid -OH groups involved in the intermolecular bond as mentioned by [66] Contreras-Padilla (2016). At 2975–2919 cm^{-1}, a band may appear which is related to the stretching of the -CH groups belonging to the pyranose groups, then a softer absorption band at 2850 cm^{-1}which is related to the stretching of $-CH_2$ groups of the carboxylic group [67]. It is shown the lack of waveband at 1749 cm^{-1} is linked to the low degree of esterification, as mentioned by [68] Rodriguez-Gonzalez et al., (2014) which indicates that the carboxyl groups are free and available to interact with water molecules and this results in their high capacity to absorb water; as well as if the free carboxyls are mixed with Ca^{2+} in the presence of water, they can form viscous structural networks. However, for O. robusta and O. atropes a slight vibration can be identified at 1730 cm^{-1}related to C=O stretching. In addition, it was found two bands at 1593 and 1388 cm^{-1} related to symmetric and asymmetric COO- stretching, which confirms the low degree of mucilage esterification. Finally, it was found a band at 1030 cm^{-1}, due to the vibration of C-O molecules attributed to the stretching of secondary cyclic alcohols [67]. The bands below 1000 cm $^{-1}$ correspond to β-D-glucose and below 800 cm $^{-1}$ are attributed to vibrations of N-H and O-H groups (**Figure 3**) [69].

5.2 Structure and general properties of pectin

Pectin is part of the structural tissues of plants and vegetables, it is present in the skin of certain fruits such as apples or in the pulp of other vegetables such as citrus fruits, strawberries, quince, and carrots. Pectin contains mainly galacturonic acid (GalA) with residues partially esterified with methanol and is the main component of the middle lamella in plants and primary cell wall, it has the function of providing cohesion and stability to tissues [70]. Pectin is formed by long chains of α-D- (1 → 4) linked to galacturonic acid interspersed by the insertion of residues (1 → 2) linked to L rhamnose residues with side chains of neutral sugars, the linear segments are predominantly composed of homogalacturonan [71]. Within its structure homogalacturonan, rhamnogalacturonan I and II, and xylogalacturonan are identified. **Figure 4** shows the basic molecular structure of pectin, where it can be seen that each ring has a carboxyl group that can be esterified with methanol-producing methyl esters [72].

The importance of pectin lies in its ability to form gels in the presence of Ca^{+2} ions or in solute at low pH, hence its importance and multiple applications as a thickening agent, gelling agent, binder, and stabilizer in industries such as pharmaceuticals for gastrointestinal treatments, in the food industry for the production of jams and frozen foods, and recently, innovations in its use for edible coatings and foams. However, for the formation of gels, the most important characteristic is the quality of the extracted pectin, which is classified into two types: pectin with a high degree of methylation and pectin with a low degree of methylation.

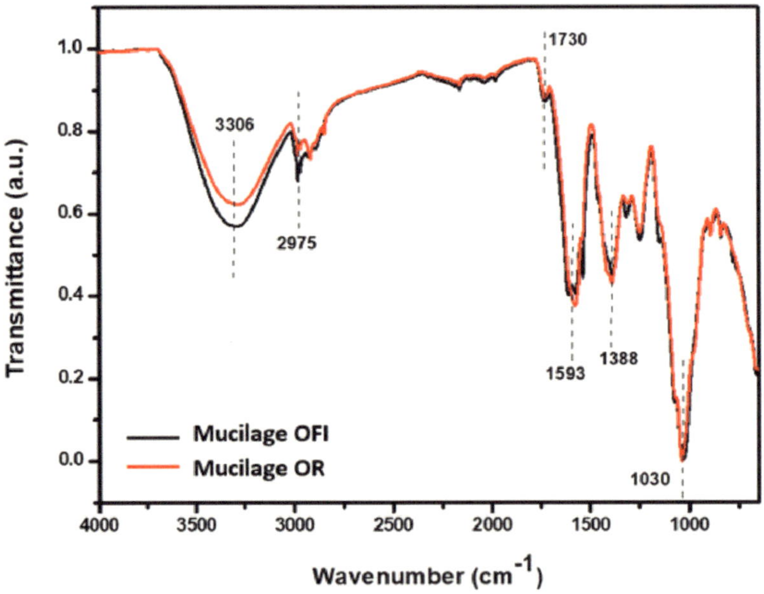

Figure 3.
ATM-FTIR mucilage extracted from Opuntia ficus-indica (OFI) and Opuntia robusta (OR) by alkaline hydrolysis.

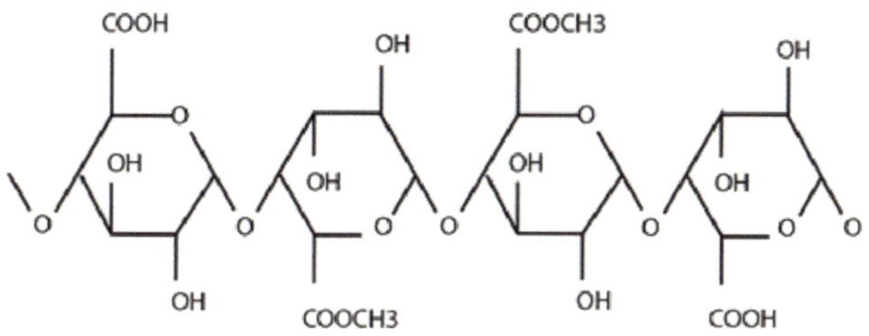

Figure 4.
Basic molecular structure of pectin [72].

The degree of methylation is the indicator of galacturonic acid residues esterified or methoxylated by the methyl group and are classified into low (<50%) and high methoxyl (>50%) pectins [73]. For low methoxyl pectin "gelation results from ionic bonding through calcium bridges between two carboxyl groups that belong to two different chains in close contact with each other. In high methoxyl pectin, cross-linking of pectin molecules involves a combination of hydrogen bonding and hydrophobic interactions between the molecules" mentions [74] Thakur et al., (2009). The yield of pectin can be from 4.42 to 10.39%, depending on conditions such as time, temperature, pH, and dry weight extraction method [75]. With the microwave-assisted extraction method, yields of 12.56% dry weight were obtained [70]. Recently, the enzymatic method with xylanase and cellulase was used for the extraction of pectin from O. ficus indica where yields of 17.91% in dry weight were obtained [76].

The conditions and method of pectin extraction directly affect the GalA content and therefore its gelation capacity; in the case of requiring pectin as a functional additive, hot acid extraction is recommended; for the use of de-esterified pectin, with

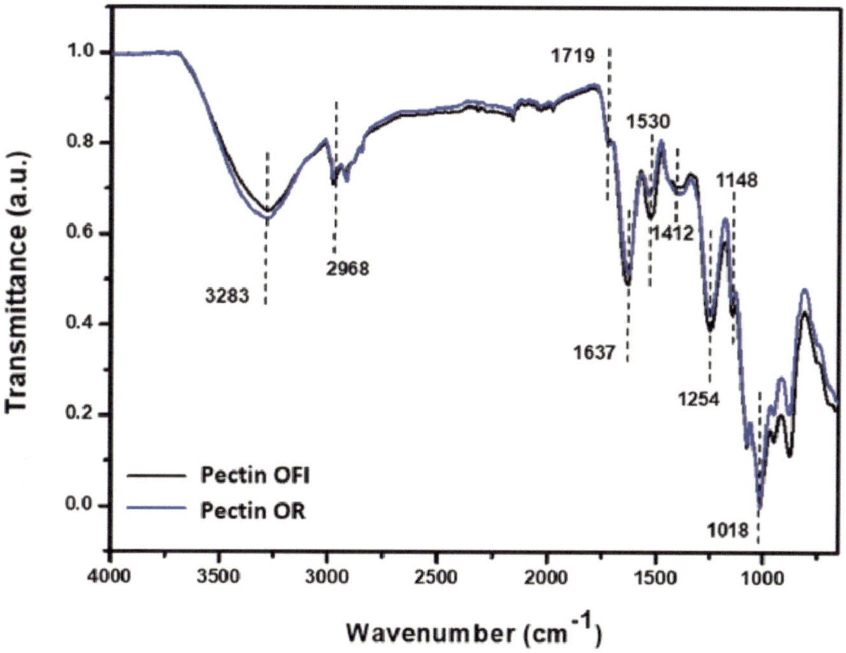

Figure 5.
ATM-FTIR of pectin extracted from Opuntia ficus-indica (OFI) and Opuntia robusta (OR) by alkaline hydrolysis.

high GalA content and gelation capacity with calcium ions, an alkaline soluble extraction is recommended [77].

According to the alkaline, extraction carried out in this study and compared to that performed by [61] Goycoolea & Cárdenas (2003), [77] Cárdenas et al., (2008) and [76] Bayar et al., (2017), the FITR analysis shows the following vibrations. A strong vibration at 3283 cm $^{-1}$ related to the stretching vibrations of the -OH groups of alcohol and carboxylic acid involved in inter-and intramolecular hydrogen bonds of the galacturonic acid polymer [78]. Subsequently at 2968 cm^{-1} a pronounced band corresponding to the absorption of the $O\text{-}CH_3$ extension bonds the methyl ester of galacturonic acid [79]. A 1720 cm^{-1} vibration caused by stretching vibration C=O methyl esterified carboxyl groups, at 1624 cm^{-1} vibration is related to the stretching of carboxylate ions and the relative ester band, which is more intense in pectins of a high degree of esterification. The bands found between 1600 to 1400 cm^{-1} correspond to the antisymmetric and COO - symmetric stretching characteristic of carboxylic acid salts. Some of the carboxyl group signals may also originate from phenolic compounds as indicated by the presence of peaks at 1530 cm^{-1} for aromatic ring vibrations [80]. It has been shown that the relative intensity of the last two peaks is related to the degree of methoxylation. The bands found between 1250 to 1140 cm^{-1} correspond to the C-O-C ether stretching [81]. The last strong bands found between 1140 to 1100 cm^{-1} are due to C-O-stretching of secondary alcohols and C-O- stretching of H in cyclic alcohols respectively [76] (**Figure 5**).

6. Conclusion

According to the analysis of the hydrocolloids extracted by acid hydrolysis contained in O. ficus-indica and O. robusta, analyzed by FTIR and compared with other research, it is concluded that the functional groups found in the mucilage are characteristic of proteins and polysaccharides, that have mechanical properties of viscosity, which can be used in industries such as food, pharmaceutical, construction, cosmetics, and biotechnology. However, when separating the solid residue from the mucilage extraction and performing acid hydrolysis to it, we obtain (according to the FTIR analysis) pectin with a low degree of methoxylation, because a small absorption peak is observed at 1732 cm^{-1}, which is attributed to the C = O stretching vibration of the carboxyl groups esterified with methyl. Due to this, the use of this pectin in combination with the indicated substances and the selected culture medium is suggested as a potential partial or total substitute for agar as a gelling agent in In vitro culture media for the development of plant cultures under laboratory conditions.

Acknowledgements

I thank the Universidad Autónoma de Querétaro and the bromatology laboratory staff of this university. To Dr. Liliana España Sánchez, research professor at the Centro de Investigación y Desarrollo Tecnológico en Electroquímica (CIDETEQ) and Dr. Reyna Araceli Mauricio Sánchez (CINVESTAV) for their assistance in the FTIR measurement and characterization performed. To my thesis advisors. Also, to the Consejo Nacional de Ciencia y Tecnología (CONACyT) for the scholarship granted to carry out this research and Call Fund for the Promotion of Entrepreneurial Culture 2021, School of Engineering, Universidad Autónoma de Querétaro.

Conflict of interest

The authors declare no conflict of interest.

Author details

Arantza Elena Sánchez-Gutiérrez[1], Genaro Martín Soto-Zarazúa[1]*, Manuel Toledano-Ayala[2] and Juan Fernando García-Trejo[1]

1 School of Engineering-Campus Amazcala, Universidad Autónoma de Querétaro, El Marqués, México

2 School of Engineering-Cerro de las campanas, Universidad Autónoma de Querétaro, Querétaro, México

*Address all correspondence to: soto_zarazua@yahoo.com.mx

References

[1] Barrows G, Sexton S, Zilberman D. Agricultural biotechnology: The promise and prospects of genetically modified crops. Journal of Economic Perspectives. 2014;**28**(1):99-120

[2] Ardisana EH, Álvarez SP, Macías RM, Gaínza BM. Future perspectives and social impact of plant biotechnologies. Alternativas. 2016;**17**(2):44-51. DOI: 10.23878/alternativas.v17i2.116

[3] Ortuño AMT, Díaz LE, Del Río JAC. Evolution of plant physiology in the last 100 years. Eubacteria Journal. 2015;**34**(2015):74-82

[4] Romay G, Matehus J, Gerstl A, Rueda R, Santana MA. Modified cassava starch as an economical substitute for solidifying agent for plant tissue culture media. Interscience. 2006;**31**(9):686-689

[5] Agrawal A, Sanayaima R, Tandon R, Tyagi RK. Cost-effective in vitro conservation of banana using alternatives of gelling agent (isabgol) and carbon source (market sugar). Acta Physiologiae Plantarum. 2010;**32**(4):703-711

[6] Gordo DAM, Gonzalez OC, Pacheco JC. Substances used as gelling agent alternatives to agar in culture media for in vitro propagation. Journal of Agricultural and Environmental Research. 2012;**3**(2):49-62

[7] López-Escamilla AL, López-Herrera M, Loaiza-Alanís C. Effect of different gelling agents on germination and in vitro development of Echinocactus platyacanthus link et otto (cactaceae) seedlings. Polybotany. 2016;**42**:153-166

[8] Jain N, Babbar SB. Gum katira-a cheap gelling agent for plant tissue culture media. Plant Cell, Tissue and Organ Culture. 2002;**71**(3):223-229

[9] Babbar SB, Jain R. Xanthan gum: An economical partial substitute for agar in microbial culture media. Current Microbiology. 2006;**52**(4):287-292

[10] Dobránszki J, Magyar-Tábori K, Tombácz E. Comparison of the rheological and diffusion properties of some gelling agents and blends and their effects on shoot multiplication. Plant Biotechnology Reports. 2011;**5**(4):345-352

[11] Martín D, Cárdenas O, Cárdenas A. Potato starch, alternative gelling agent in culture media for in-vitro propagation of lulo Solanum quitoense lam. The Journal of Agricultural Science. 2013;**30**(1):3-11

[12] Gallegos-Villela RR, Larrea-Zambrano FD, Goyes-Lopez CE, Perez-Sanchez JF, Suarez-Dominguez EJ, Palacio-Perez A. Effect of natural additives on concrete mechanical properties. Cogent Engineering. 2021;**8**(1):1870790

[13] Madera-Santana TJ, Vargas-Rodríguez L, Núñez-Colín CA, González-García G, Peña-Caballero V, Núñez-Gastélum JA, et al. Mucilage from cladodes of Opuntia spinulifera Salm-Dyck: Chemical, morphological, structural and thermal characterization. CyTA Journal of Food. 2018;**16**(1):650-657

[14] Mascot-Gómez E, Flores J, López-Lozano NE, Yáñez-Espinosa L. Seed germination of southern Chihuahua desert cacti: Effect of mucilage, light and phytohormones. Flora. 2020;**263**:151528. DOI: 10.1016/j.flora.2019.151528

[15] Villa-Uvidia DN, Osorio-Rivera MÁ, Villacis-Venegas NY. Extraction, properties and benefits of mucilages. Science Domain. 2020;**6**(2):503-524

[16] Calva GC, Pérez JV. Plant cell and tissue culture: Source of food for the future. Revista Digital Universitaria. 2005;**6**(11):1-16

[17] Vasil IK. A short history of plant biotechnology. Phytochemistry Reviews. 2008;**7**:387-394. DOI: 10.1007/s11101-007-9075-z

[18] Gautheret RJ. Plant tissue culture: A history. The Botanical Magazine. 1983;**96**:393-410. DOI: 10.1007/BF02488184

[19] Murashige T, Skoog F. A revised medium for rapid growth and bio assays with tobaoco tissue cultures. Physiologia Plantarum. 1962;**15**:1-26

[20] Richard L, Jarret ERL. Plant regeneration in tissue culture: Somatic embryogenesis and organogenesis. In: Tissue Culture in Agriculture: Fundamentals and Applications. International Center for Tropical Agriculture (CIAT); 1991. pp. 143-172. ISBN 958-9183-15-8. Available from: http://ciat-library.ciat.cgiar.org/Articulos_Ciat/biblioteca/Cultivo_de_tejidos_en_la_agricultura.pdf

[21] Gamborg OL, Murashige T, Thorpe TA, Vasil IK. Plant tissue culture media. In Vitro. 1976;**13**(7):473-479

[22] Phillips GC, Garda M. Plant tissue culture media and practices: An overview. In Vitro Cellular & Developmental Biology. Plant. 2019;**55**:242-257. DOI: 10.1007/s11627-019-09983-5

[23] Saad AIMM, Elshahe AM. Plant Tissue Culture Media. In: Leva A, Rinald LMR, editor. Recent Advances in Plant in vitro Culture. IntechOpen; 2012. DOI: 10.5772/50569. Available from: https://www.intechopen.com/chapters/40181

[24] Roca WM, Mroginski LA. Tissue Culture in Agriculture: Fundamentals and Applications. International Center for Tropical Agriculture CIAT. Chap. 2; 1991. p. 32. Available from: http://ciat-library.ciat.cgiar.org/Articulos_Ciat/biblioteca/Cultivo_de_tejidos_en_la_agricultura.pdf

[25] Padmesh S, Singh A. Agars: Properties and applications. In: Polysaccharides: Properties and Applications. Wiley. Chapter 5; 2021. pp. 75-94. DOI: 10.1002/9781119711414.ch5

[26] Araki C. Structure of the agarose constituent of agar-agar. Short Communication. Faculty of Industrial Arts Kyoto Technical University Matsugasaki, Kyoto. 1956;**29**(4):543-544

[27] Mouradi-Givernaud A, Hassani LA, Givernaud T, Lemione Y, Benharbet O. Biology and agar composition of Gelidium sesquipedale harvested along the Atlantic coast of Maroco. Carbohydrate Polymers. 1999;**399**:391-395

[28] Sánchez MV, Corella R. Extraction, identification and microbiological testing of agar extracted from Gracilaria fortissima dawson (Rhodophyta, Gigartinales, Gracilariaceae). Uniciencia. 2008;**22**(1-2):99-106

[29] Aram C, Hirase S. Exhaustive Mercaptolyses of agar-agar. Studies on the Chemical Constitution of Agar-agar XV. 1953;**26**(8):463-467

[30] Selvalakshmi S, Vijaya N, Selvasekarapandian S, Premalatha M. Biopolymer agar-agar doped with NH_4SCN as solid polymer electrolyte for electrochemical cell application. Journal of Applied Polymer Science. 2017;**134**(15):1-10. DOI: 10.1002/app.44702

[31] Stancioff DJ, Stanley NF. Infrared and chemical studies of algal

polysaccharides. Proceedings of the International Seaweed Symposium. 1969;**6**:595-609

[32] Guerrero P, Etxabide A, Leceta I, Peñalba M, De la Caba K. Extraction of agar from Gelidium sesquipedale (Rodhopyta) and surface characterization of agar based films. Carbohydrate Polymers. 2014;**99**:491-498

[33] El-hefian EA, Mahmoud MN, Hamid A, Y. Preparation and characterization of chitosan/agar blended films: Part 1. Chemical structure and morphology. Journal of Chemistry. 2012;**9**(3):1431-1439

[34] Montilla-Escudero EA, Dulce-Rivadeneira MF, Quevedo-Hidalgo B, Mercado-Reyes M, Álvarez-León R, Molina-Vargas JN, et al. Effect of alkaline treatment on the productivity and physical properties of agar-agar from Gracilaria verrucosa. Marine and Coastal Research Bulletin. 2011;**40**(1):75-88

[35] Freile-Pelegrín Y, Robledo D, Pedersén M, Bruno E, Rönnqvist J. Effect of dark and salinity treatment on the yield and agar quality of Gracilaria cornea (Rhodophyceae). Marine Sciences. 2002;**28**(3):289-296

[36] Macler BA, West JA. Life history and physiology of the red alga, Gelidium coulteri, in unialgal culture. Aquaculture. 1987;**61**(3-4):281-293

[37] Hurtado MA, Manzano-Sarabia M, Hernández-Garibay E, Pacheco-Ruíz I, Zertuche-González JA. Latitudinal variations of the yield and quality of agar from Gelidium robustum (Gelidiales, Rhodophyta) from the main commercial harvest beds along the western coast of the Baja California peninsula, Mexico. Journal of Applied Phycology. 2010;**23**:727-734

[38] Mouradi-Givernaud A, Amina Hassani L, Givernaud T, Lemoine Y, Benharbet O. Biology and agar composition of Gelidium sesquipedale harvested along the Atlantic coast of Morocco. Hydrobiologia. 1999;**398**:391-395. DOI: 10.1023/A:1017094231494

[39] Zecchinel EF, Brito LL, Lárez GE. Physical properties of agar from Gracilariopsis tenuifrons (Gracilariaceae) in Sucre, Venezuela. Journal of Tropical Biology. 2000;**48**(1):193-197

[40] Kumar V, Fotedar R. Agar extraction process for Gracilaria cliftonii (Withell, Millar, & Kraft, 1994). Carbohydrate Polymers. 2009;**78**(4):813-819. DOI: 10.1016/j.carbpol.2009.07.001

[41] Chen H, Xiao Q, Weng H, Zhang Y, Yang Q, Xiao A. Extraction of sulfated agar from Gracilaria lemaneiformis using hydrogen peroxide-assisted enzymatic method. Carbohydrate Polymers. 2020;**232**:115790. DOI: 10.1016/j.carbpol. 2019.115790

[42] Porse H, Rudolph B. The seaweed hydrocolloid industry: 2016 updates, requirements, and outlook. Journal of Applied Phycology. 2016;**29**:2187-2200. DOI: 10.1007/s10811-017-1144-0

[43] McHugh, DJ. Prospects for seaweed production in developing countries. FAO Fisheries Circular. No. 968. Rome, FAO. 2002. 28p. Available from: https://www.fao.org/3/y3550e/Y3550E01.htm

[44] Ichi T, Koda T, Asai I, Hatanaka A, Sekiya J. Effects of gelling agents on in vitro culture of plant tissues. Agricultural and Biological Chemistry. 1986;**50**(9): 2397-2399

[45] Rodríguez HG, Hechevarría IS. Aloe vera (L.) NL Burm. Gel and sago flour as solid support of culture medium for medicinal plants. Cuban Journal

of Medicinal Plants. 2006;**11**(1). Available from: http://scielo.sld.cu/scielo.php?script=sci_arttext&pid=s1028-47962006000100007

[46] Soler NP, Abril JLB. Aloe vera (aloe Barbadensis miller) in the generation of agraz (Vaccinium Meridionale Swartz) explants. Scientific Culture. 2016;**14**:58-68

[47] Mengesha A, Ayenew B, Gebremariam A, Tadesse T. Micro-propagation of Vanilla planifolia using Enset (Ensete ventricosum (Welw, cheesman)) starch as a gelling agent. Journal of Biological Sciences. 2012;**4**(4):519-525

[48] Martín D, Cárdenas O, Cárdenas A. Potato starch, alternative gelling agent in culture media for in-vitro propagation of lulo Solanum quitoense lam. The Journal of Agricultural Science. 2013;**30**(1):3-11

[49] Pineda OP, Oyuela OW. Synopsis of the genus Opuntia mill. (Opuntioideae, Cactaceae) in Honduras. Revista Forestal del Peru. 2020;**35**(1):54-72

[50] Ahmed SN, Ahmad M, Zafar M, Rashid S, Sultana S. Classification, distribution and morphological characterization of Opuntia species. In: Ramadan MF, Tem A, Rohn S, editors. Opuntia Spp. Chemistry, Bioactivity and Industrial Cham: Applications. Springer; 2021. DOI: 10.1007/978-3-030-78444-7_5

[51] Salehi E, Emam-Djomeh Z, Askari G, Fathi M. Opuntia ficus-indica fruit gum: Extraction, characterization, antioxidant activity and functional properties. Carbohydrate Polymers. 2019;**206**:565-572. DOI: 10.1016/j.carbpol.2018.11.035

[52] Hernández-Urbiola MI, Contreras-Padilla M, Pérez-Torrero E, Hernández-Quevedo G, Rojas-Molina JI, Cortes ME, et al. Study of nutritional composition of Nopal (Opuntia ficus-indica cv. Redonda) at different maturity stages. The Open Nutrition Journal. 2010;**4**:11-16

[53] Bensadón S, Hervert-Hernández D, Sáyago-Ayerdi SG, Goñi I. By-products of Opuntia ficus-indica as a source of antioxidant dietary Fiber. Plant Foods for Human Nutrition. 2010;**65**:210-216. DOI: 10.1007/s11130-010-0176-2

[54] Remes-Troche JM, Taboada-Liceaga H, Gill S, Amieva-Balmori M, Rossi M, Hernández-Ramírez G, et al. Nopal fiber (Opuntia ficus-indica) improves symptoms in irritable bowel syndrome in the short term: A randomized controlled trial. Neurogastroenterology and Motility. 2021;**33**(2):e13986. DOI: 10.1111/nmo.13986

[55] Di Lorenzo F, Silipo A, Molinaro A, Parrilli M, Schiraldi C, D'Agostino A, et al. The polysaccharide and low molecular weight components of Opuntia ficus-indica cladodes: Structure and skin repairing properties. Carbohydrate Polymers. 2017;**157**:128-136. DOI: 10.1016/j.carbpol.2016.09.073

[56] Henríquez-Rodríguez MA, Pérez J, Gascó J, Rodríguez O, Prieto A. Preliminary biochemical characterization of the main components from the Cardon. Journal Unell. Sci. Tec. 2010;**27**:95-102

[57] Sepúlveda E, Sáenz C, Aliaga E, Aceituno C. Extraction and characterization of mucilage in Opuntia spp. Journal of Arid Environments. 2007;**68**(4):534-545. DOI: 10.1016/j.jaridenv.2006.08.001

[58] Trachtenberg S, Mayer M. Mucilage cells, calcium oxalate crystals and soluble calcium in Opuntia ficus-indica. Annals of Botany Company. 1982;**50**:549-557

[59] Saag LMK, Sanderson GR, Moyna P, Ramos G. Cactaceae mucilage composition.

Journal of the Science of Food and Agriculture. 1975;**26**(7):993-1000

[60] Medina-Torres L, Brito-De la Fuente E, Torrestiana-Sánchez B, Katthain R. Rheological properties of the mucilage gum (Opuntia ficus-indica). Food Hydrocolloids. 2000;**14**(5):417-424. DOI: 10.1016/S0268-005X(00)00015-1

[61] Goycoolea FM, Cárdenas A. Pectins from Opuntia spp.: A short review. Journal of the Professional Association for Cactus Development. 2003;**5**(1):17-29

[62] Sáenz C, Sepúlveda E, Matsuhirob B. Opuntia spp. mucilage's: A functional component with industrial perspectives. Journal of Arid Environments. 2004;**54**:275-290. DOI: 10.1016/S0140-1963(03)00106-X

[63] McGarvie D, Parolis H. Methylation analysis of the mucilage of Opuntia ficus-indica. Carbohydrate Research. 1981;**88**:305-334

[64] Sáenz C, García N, Abraján M, Fabry AM, Robert P. The nopal cactus, an arid zone species producer of natural hydrocolloids. Latin American Food. 2016;**50**(322):60-68

[65] Majdoub H, Roudesli S, Picton L, Le Cerf D, Muller G, Grisel M. Prickly pear nopal pectin from Opuntia ficus-indica physico-chemical study in dilute and semi-dilute solutions. Carbohydrate Polymers. 2001;**46**(1):69-79

[66] Contreras-Padilla M, Rodríguez-García ME, Gutiérrez-Cortez E, Valderrama-Bravo MC, Rojas-Molina JI, Rivera-Muñoz EM. Physicochemical and rheological characterization of Opuntia ficus mucilage at three different maturity stages of cladode. European Polymer Journal. 2016;**78**:226-234. DOI: 10.1016

[67] Monrroy M, García E, Ríos K, García JR. Extraction and physicochemical characterization of mucilage from Opuntia cochenillifera (L.) miller. Journal of Chemistry. 2017;**2017**:1-9. DOI: 10.1155/2017/4301901

[68] Rodríguez-González S, Martínez-Flores HE, Chávez-Moreno CK, Macías-Rodríguez LI, Zavala-Mendoza E, Garnica-Romo MG, et al. Extraction and characterization of mucilage from wild species of Opuntia. Journal of Food Process Engineering. 2014;**37**:285-292. DOI: 10.1111/jfpe.12084

[69] Zhao M, Yang N, Yang B, Jiang Y, Zhang G. Structural characterization of water-soluble polysaccharides from Opuntia monacantha cladodes in relation to their anti-glycated activities. Food Chemistry. 2007;**105**(2007):1480-1486. DOI: 10.1016/j.foodchem.2007.05.031

[70] Lefsih K, Giacomazza D, Dahmoune F, Mangione MR, Bulone D, San Biagio PR, et al. Pectin from Opuntia ficus-indica: Optimization of microwave-assisted extraction and preliminary characterization. Food Chemistry. 2016;**221**:91-99. DOI: 10.1016/j.foodchem.2016.10.073

[71] Martínez-González CR, Ramírez-Mendoza R, Jiménez-Ramírez J, Gallegos-Vázquez C, Luna-Vega I. Improved method for genomic DNA extraction for Opuntia mill. (Cactaceae). Plant Methods. 2017;**13**(82):1-10. DOI: 10.1186/s13007-017-0234-y

[72] Zegada VYF. Extraction of pectin from orange peel waste by microwave-assisted acid hydrolysis (MOH). Research and Development. 2015;**1**(15):65-76

[73] Mendoza-Vargas L, Jiménez-Forero J, Ramírez-Niño M. Evaluation of enzymatically extracted pectin from cocoa (Theobroma cacao L.)

fruit shells. Revista U.D.C.A Actualidad & Divulgación Científica. 2017;**20**(1):131-138

[74] Thakur BR, Singh RK, Handa AK, Rao MA. Chemistry and uses of pectin - a review. Critical Reviews in Food Science and Nutrition. 2009;**37**:47-73. DOI: 10.1080/10408399709527767

[75] Pérez-Martínez JD, Sánchez-Becerril M, Ornelas-Paz JJ, González-Chávez MM, Ibarra-Junquera V, Escalante-Minakata P. The effect of extraction conditions on the chemical characteristics of pectin from Opuntia ficus-indica cladode flour. Journal Polymer Enviroment. 2013;**21**(2013):1040-1051. DOI: 10.1007/s10924-013-0602-2

[76] Bayar N, Friji M, Kammoun R. Optimization of enzymatic extraction of pectin from Opuntia ficus-indica cladodes after mucilage removal. Food Chemestry. 2017;**241**:127-134. DOI: 10.1016/j.foodchem.2017.08.051

[77] Cárdenas A, Goycoolea F, Rinaudo M. On the gelling behaviour of 'nopal' (Opuntia ficus-indica) low methoxyl pectin. Carbohydrate Polymers. 2008;**73**(2008):212-222. DOI: 10.1016/j.carbpol.2007.11.017

[78] Santos EEE, Amaro RC, Bustamante CCCC, Guerra MHA, Soares LC, Froes RES. Extraction of pectin from agroindustrial residue with an ecofriendly solvent: Use of FTIR and chemometrics to differentiate pectins according to degree of methyl esterification. Food Hydrocolloids. 2020;**107**(2020):105921. DOI: 10.1016/j.foodhyd.2020.105921

[79] Gopi D, Kanimozhi K, Kavitha L. Opuntia ficus-indica peel derived pectin mediated hydroxyapatite nanoparticles: Synthesis, spectral characterization, biological and antimicrobial activities. Spectrochimica Acta Part A: Molecular and Biomolecular Spectroscopy. 2015;**141**:135-143. DOI: 10.1016/j.saa.2015.01.039

[80] Wang W, Ma X, Xu Y, Cao Y, Jiang Z, Ding T, et al. Ultrasound-assisted heating extraction of pectin from grapefruit peel: Optimization and comparison with the conventional method. Food Chemistry. 2015;**178**:106-114. DOI: 10.1016/j.foodchem.2015.01.080

[81] Canteri MH, Moreno L, Wosiacki G, Scheer ADP. Pectin: From raw material to the final product. Polymers. 2012;**22**(2):149-157

Chapter 4

Soilless Culture for Agribusiness throughout Urban Farming in Indonesia

Ristina Siti Sundari, Lies Sulistyowati, Trisna Insan Noor and Iwan Setiawan

Abstract

Urban farming is proliferated worldwide related to large urban areas, and population in the city and people need food closer and healthier. Limited area forced urban dwellers to grow food on landless use or soilless culture to cultivate any agricultural commodities by the citizen in the urban area. An urban farmer is not a real farmer. Urban farmers are interested in growing plants or fish and have limited farming skills, but their curiosity becomes essential for urban farming activities. Urban farming has become more beneficial edible food that starts from interesting in the environment, green city, and planting hobby, but the soil and land are limited despite prohibited land scarcity and soil medium. However, urban farming used soilless culture for rooting medium. The soilless medium included less or minimum soil, hydroponics, aquaponics, aeroponic. Soilless culture is now set as vertical farming, particularly in the urban area with finite land. Urban farming as an agribusiness is dominant for household self-demand than commercial. In the future urban farming, productivity can feed the city.

Keywords: agribusiness, edible plant, urban farming

1. Introduction

Soilless culture is the cultivation system in the agriculture sector that uses less soil medium using another medium to grow the plant or fish. Soilless culture develops fast along with population growth and land conversion, whether we want it or not, planting edible food to feed, herbs, and medicines for urban population need to be sustained.

There have been dramatic changes in this, even along with pandemic hits. Urban soilless farming is a feasible production strategy for many plants farmed as specialized foods, herbs, or medicinal plants. Vertical farms have grown in popularity to integrate modern production systems into cities [1]. While there are various obstacles, it is apparent that this is a fast-increasing field of urban farming. The science of plant development in soilless systems is still in its infancy, and despite significant progress, many concerns remain unresolved. One of the goals of this book is to focus on the central issues of urban farming in soilless culture. Future research is needed to take full advantage of the economic, social, and ecological aspects to sustain. Typical soilless media are generally described by root limitation, restricted substrate volume, insufficient solution composition buffer capacity, and limited nutrient input. Commercially, substrate culture in plastic bags has been used effectively for fruiting crops. In deciding on a soilless culture method, the production size, crops, growth cycles and length of each cycle, cost, and managerial abilities, as well as environmental issues are examined. As a result, closed loop technologies have grown in importance to boost sustainability in soilless agriculture. The purpose of this presentation is to describe several soilless growing techniques and their potential for application in vegetable production [2]. Compared with soil-based production, soilless cultivation of plants in containers requires a limited root system and a lower root zone volume. It is vital to note that under these conditions, the physical and chemical qualities of the growth medium must be such that the plant can store enough water and nutrients while maintaining enough aeration. Plants in pots and containers were traditionally grown using resources such as locally available soils. Organic elements such as manures and composts were frequently utilized to augment the inorganic foundation materials, enhancing water retention and airspace while also providing nutritional advantages. Growing media used commercially in many regions of the globe in the early twenty-first century are primarily made from peat and other organic materials, with the notable exceptions being mineral wool substrates used for glasshouse crops such as tomatoes, cucumbers, and pepper [3].

The critical need is market demand for a particular product, differentiating between common greenhouse crops and other minor crops or regionally consumed items. In all circumstances, crop selection should provide an optimal fit between production and market delivery time, given rising production costs and the limited shelf-life of vegetable goods. Economic issues are concerned with the link between market pricing and farmer returns [4]. Soilless cultivation is used in almost all urban farming. The specifics will be detailed further below.

2. Urban farming activity by soilless medium

Almost urban farming carried out in West Java urban area cultivated in limited land, limited soil, and soilless medium depends on the capital they have, generally influenced by both horticultural and economic factors. This aspect is ultimately the primary motivator for designing a specific substrate mix that is used in a soilless production scenario. The financial aspect presents itself in the availability of supplies, processing costs, transportation costs, costs related to plant/crop production, and transportation and marketing. In some circumstances, the disposal of spent substrates is a crucial matter with environmental and economic significance [3]. Some are from the government grant program to secure community food.

Growers believe peat to be less expensive in nations where it is widely available, maybe even collected locally, than in countries where it must be imported from faraway regions. With raw material costs fluctuating, producers must decide whether to employ a "tried and true" component such as peat or a substitute [3] such as cocopeat, husks, and sawdust in a mix that has proved to perform well over the year. In some years, the financial circumstances may cause a modification to be considered. Because the qualities of all substrates and mixtures vary, replacing one component as cocopeat or peat with another may result in more expenses or inferior quality crops that have

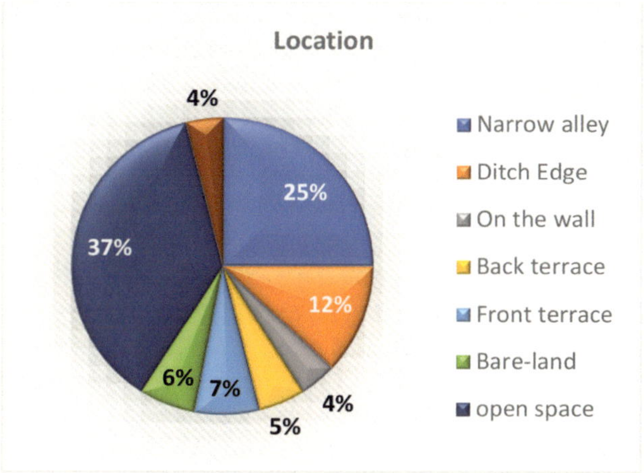

Figure 1.
Soilless urban farming location.

less value in the market, especially if the substitution is a material with which the producer has less expertise. Growers worldwide are faced with the task of building blends that will show as intended while costing the least amount of money overall.

Vertical farming is a popular method in urban farming, growing in a tight and narrow alley, along the side of a ditch, on a wall, in a hanging pot, on a rooftop, terrace, bareland, or in open space is an example of urban farming.

The research in Tasikmalaya and Banjar City found that urban farming in the open space is still dominant (37%) and followed respectively by narrow (tight) alley, ditch edge (12%), front terrace (7%), bare-land (6%), back terrace (5%), wall and rooftop (4%). On the household scale, there were no cultivated plants or fish in Greenhouse. It is due to not yet commercial produce (**Figure 1**).

While used technology in soilless culture in pot medium was dominant (69%), followed by conventional (19%), hydroponics (7%), aquaponics (4%), and aeroponics (1%).

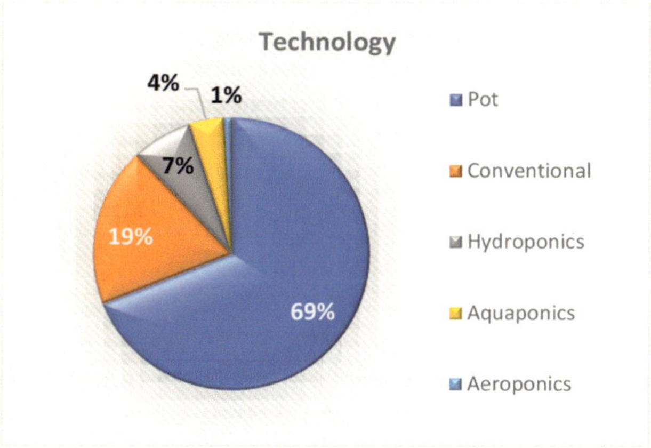

Figure 2.
Soilless urban farming technology.

The irrigation can be used in various systems such as: sprinkler, direct watering, drip irrigation, pool, from the ditch, etc. The soilless culture in an urban farming medium often uses less soil, cocopeat, peat, sawdust, rockwool, hydroleca/hydro tone, charcoal, perlite, hydrogel, vermiculite, zeolite (**Figure 2**).

Soilless culture is not separated by applied technology as follows:

2.1 Pot medium

Growing plants with less soil then mix with compost, burned husks, cocopeat, sawdust, etc. World agriculture has changed considerably in the previous several decades, and this transformation will continue as long as the underlying reasons for these changes exist. The fast scientific, economic, and technological growth of societies throughout the world is one of these drivers. The world's population is growing, and many countries' living standards are improving. With improved living standards, there has been significant demand for floricultural crops such as cut flowers, pot plants, and bedding plants. As a result of these tendencies, the usage of a large range of covered agricultural systems, ranging from crude screen or plastic film coverings to fully climate-controlled greenhouses, had increased. Initially, this manufacturing took place in the ground, where the soil had been changed to allow for adequate drainage. Because the production costs of protected culture are greater than those of open field cultivation, growers were forced to raise their production intensity to remain competitive. Plant production was accomplished through various strategies, the most notable of which was the fast expansion of soilless output compared with overall agricultural crop production. Most such plants are grown in greenhouses, generally under near-optimal production conditions on a commercial scale. An inherent drawback of soilless cultivation over soil-based cultivation is that the root volume is unrestricted in the latter, while in containerized culture, the root volume is restricted. This restricted root volume has several significant effects, especially a limited supply of water [1] and nutrients [2]. The limited root volume also increases root-to-root competition since the medium has more roots per unit volume.

Container production systems have advantages over in-ground production systems such as pollution prevention since it is possible, and these growing systems are used for minimizing or eliminating the discharge of nutritional ions and pesticide residues. However, freshwater reservoirs should be conserved. Water and nutrient have been used efficiently that are significantly greater in container production and have clear economic benefits. More and more attention is being directed to reducing the environmental pollution especially in the developed country. In the countries, this type of production represents a large portion of agricultural productivity. Runoff from nursery and greenhouse must be minimized or eliminated by creating regulations to force recirculation from the nurseries and greenhouses (**Figure 3**).

2.2 Hydroponics

Growing plant without soil: The process of growing plants in nutrient solutions is known as hydroponics. Furthermore, for systems that employ just nutrient solution and air such as deep flow technique (DFT), nutrient film technique (NFT), and aero-hydroponics, we include in this notion substrate-based systems in which the substrate does not provide nutrients or ionic adsorption or exchange. As a result, we define hydroponic production systems to use inert substrates such as rockwool or gravel. The term "hydroponics" is in conjunction with qualifying terms to clarify the

DOI: http://dx.doi.org/10.5772/intechopen.101757

Figure 3.
Urban farming throughout pot medium technology.

distinction between "liquid-culture hydroponics" and "substrate-based hydroponics"; The latter involves a significant quantity of inert substrate with limited ion exchange capability, whereas the former contains none (or virtually none) of any substrate. Initially, scientists primarily employed hydroponics as a research tool to explore some aspects of plant nutrition and root function. Progress in plastics manufacture, automation, the fabrication of totally soluble fertilizers, and, most importantly, the development of many different substrates supplemented scientific breakthroughs and pushed soilless farming to a commercially feasible level. There are several types of soilless systems available today for producing vegetables and ornamentals in greenhouses. As a result, there is a vast range of growth systems. The substrate for hydroponics plants commonly uses hydroton or hydroleca, rockwool, perlite, vermiculite, and cocopeat (**Figure 4**) [3, 4].

2.3 Aeroponics

Growing plant without soil: Cultivating plants in an opaque that supports container in which their roots are suspended and showered in a nourishing mist is known as aeroponics. The type of the aeroponic, frequency of root exposure, and content of the nutritional solution are the essential features of the approach. Spraying typically lasts 30 until 60 seconds, and the frequency varies depending on the species, plant development stage, growing season, and time of day. The drainage from each nebulization is collected at the bottom of the modules and recirculated [5]. The aeroponic system also gave unique insights into storage root development by triggering the proliferation of auxin-dependent towards secondary xylem parenchyma cells to generate early root thickness and bulk. The created approach can aid molecular scientists, breeders, and physiologists directly throughout, allowing them to screen germplasm on root qualities that correspond with enhanced economic attributes (**Figure 5**) [6].

Figure 4.
Urban farming throughout hydroponics technology.

Figure 5.
Urban farming throughout aeroponics technology.

2.4 Aquaponics

Growing plant without soil: The demands to increase food production expand, and as the stresses on resources such as water, land, and nutrients become more significant when the world's population grows. There is an urgent need to find alternative, sustainable, and reliable methods to provide this food [7, 8]. The current strategies for supplying more produce are neither ecologically sound nor address the issues of the circular economy of reducing waste while meeting the WHO's Millennium Development Goals of zero hunger and poverty by 2015 [9]. Aquaponics is an aquaculture and hydroponics technology integration provides to reset the solution. Aquaponics has been identified as a farming approach that, through nutrient and waste recycling, can aid in addressing both planetary boundaries and sustainable development goals, particularly for arid regions or areas with nonarable soils [1, 10, 11]. Due to aquaponics technology needing more capital, some priority strategies such as upgrading any skill and knowledge deal with aquaponics, socializing system benefit, increasing aquaponics yield consumption and consumer expendable, implementing the food health by aquaponics system, and upgrading innovative agribusiness capacity to enhance distribution access and procure

Figure 6.
Urban farming throughout aquaponics.

Figure 7.
Urban farming through anoponics.

lower start-up cost [12]. The simple aquaponics in Indonesia is called *Budikdamber* that means planting aquaponics in the bucket (**Figure 6**).

2.5 Anoponiccs

Growing plants without anything: neither medium nor nutrition we give. It is considered from air fixation to get nutrition. Spanish moss, for example, grows well in Indonesia. The ornament plant is not yet well known. So it needs further research about mechanism, metabolism and others related to it (**Figure** 7).

3. Conclusions

Soilless culture is non-traditional farming. Soilless culture leads to urban farming that is based on capital, skill, and technology. Nevertheless, growing plant or fish under soilless culture gives much higher productivity even not for commercial. Soilless in urban farming is fast growing to provide food from home. The technologies used were pot medium, hydroponics, aquaponics, aeroponics. Soilless culture may be in various ways such as tight alley, along the side of a ditch, on a wall, in a hanging pot, on a rooftop, terrace, bare ground. We found the plant cultivation that grows without any medium, such as Spanish Moss.

Conflict of interest

The authors declare no conflict of interest.

Author details

Ristina Siti Sundari[1,2]*, Lies Sulistyowati[2], Trisna Insan Noor[2] and Iwan Setiawan[2]

1 University of Perjuangan, Tasikmalaya, Indonesia

2 University of Padjdjaran, Bandung, Indonesia

*Address all correspondence to: ristina.sitisundari@yahoo.com

References

[1] König B, Junge R, Bittsanszky A, Villarroel M, Komives T. On the sustainability of aquaponics. Ecocyles. 2016;**2**(1):26-32. DOI: 10.19040/ecocycles.v2i1.50

[2] Dubik S, Krizek DT, Smart DP. Influence of root zone restriction on mineral element concentration, water potential, chlorophyll concentration, and partitioning of assimilating in spreading euonymus (e. kiautschovica loes. 'sieboldiana'). Journal of Plant Nutrition. 1990;**13**(6):677-699. DOI: 10.1080/01904169009364109

[3] Khandaker M, Kotzen B. The potential for combining living wall and vertical farming systems with aquaponics with special emphasis on substrates. Aquaculture Research. 2018;**49**(4):1454-1468. DOI: 10.1111/are.13601

[4] Proksch G, Ianchenko A, Kotzen B. Aquaponics Food Production Systems, Combined Aquaculture and Hydroponic Production Technologies for the Future. Switzerland: Springer Nature; July 2019

[5] Savvas D, Gruda N. Application of soilless culture technologies in the modern greenhouse industry—A review. European Journal of Horticultural Science. 2018;**83**(5):280-293. DOI: 10.17660/eJHS.2018/83.5.2

[6] Selvaraj MG, Montoya-P ME, Atanbori J, French AP, Pridmore T. A low-cost aeroponic phenotyping system for storage root development: Unravelling the below-ground secrets of cassava (Manihot esculenta). Plant Methods. 2019;**15**(1):1-13. DOI: 10.1186/s13007-019-0517-6

[7] dos Santos MJPL. Smart cities and urban areas—Aquaponics as innovative urban agriculture. Urban Forestry & Urban Greening. 2016;**20**:402-406. DOI: 10.1016/j.ufug.2016.10.004

[8] Rinehart BL. "Aquaponics – Multitrophic Systems for Sustainable Food Production," NCAT. ATTRA Sustainable Agriculture; March 2019

[9] Wahyuningsih W. Millenium develompent goals (Mdgs) Dan sustainable development goals (Sdgs) Dalam Kesejahteraan Sosial. Bisma. 2018;**11**(3):390. DOI: 10.19184/bisma.v11i3.6479

[10] Khandaker M, Kotzen B. The potential for combining living wall and vertical farming systems with aquaponics with special emphasis on substrates. Aquaculture Research. January 2018;**13601**:1-15. DOI: 10.1111/are.13601

[11] Goddek S, Joyce A, Kotzen B. Aquaponics Food Production Systems, Combined Aquaculture and Hydroponic Production Technologies for the Future. Switzerland: Springer Nature; June 2019

[12] Sundari RS, Arshad A, Sulistyowati L, Noor TI, Setiawan I. Enhancing food security throughout Aquaponicsin urban farming development strategy. Journal of Physics: Conference Series. 2021;**1764**(1):122091. DOI: 10.1088/1742-6596/1764/1/012209

Chapter 5

Obtaining Cell Cultures of Medicinal Plants

Torkwase Emmanuella Bulya, Tatiana V. Glukhareva and Elena G. Kovaleva

Abstract

In vitro propagation of medicinal plants has been incorporated into producing healthy plants that are beneficial to humanity. Some basic principles and factors tend to influence the cultivation process, thus, causing this method of plant propagation to be adapted owing to the importance and benefits surrounding this method. The main objective of this research work was to obtain cell cultures of medicinal plants of *Cichorium intybus, Stevia rebaudiana Bertoni, Monarda citriodora, and Rhodiola krylovii*. In obtaining the cell cultures of these medicinal plants, some steps need to be followed. In this research, the effect of different methods of sterilisation/cultivation of plant seeds and explants were evaluated using two different media compositions, observable differences between sterile and non-sterile plant seedlings of *C. intybus, Monarda citriodora, and Rhodiola krylovii.* The effect of growth regulator (Kinetin) and non-growth regulator (Kinetin) on the cell cultures was observed in solid and liquid media; the dry and wet weight was determined for a callus of Chicory grown in cell suspension culture. All results were presented on tables and charts.

Keywords: medicinal plants, callus, solid media, liquid media, suspension culture, *Cichorium intybus*, *Stevia rebaudiana Bertoni*, *Monarda citriodora*, *Rhodiola krylovii*

1. Introduction

Plant cell cultivation based on the *in vitro* process is not general for medicinal species. Therefore, many questions are yet to be dealt with regarding the cultivation and quality assessment of the plants produced via the *in vitro* method of cultivating these plant materials, such as seeds, callus, seedlings, and hairy roots. Plant cell culture is a technology that investigates some of the conditions that promote cell division and other growth factors in *in vitro* conditions, and it is considered a valuable tool in both primary and applied studies and commercial applications [1, 2]. It has been demonstrated in previous studies that factors, such as irrigation and nutrient status, affect the chemical profile and composition of plants. Since the understanding is limited of the mechanisms and activity of herbal formulations to combat disease, quality assessment is often reduced to quantifying one or a few compounds. Herbal formulations are, however, very complex, and only a few selected compounds, can be optimal to determine the changes in the chemical profile and composition of cultivated plants [3].

Medicinal plant cultivation employs basic materials, such as substrate, small mineral elements, water, and light, to grow without stress compared to animal rearing. Environmental conditions and climatic conditions can harm plants growing in an outdoor setting. Despite the changes caused by environmental factors, such as elevated temperature, high humidity, and others, thus, to avoid repeatability and uniformity across the experiment, it is necessary to define and control specific growth conditions by growing plants indoors for research [4].

The progressiveness of subsistence culturing of plants used for commercial trading in the area of medicinal plants has assumed rise to an increase in the rate at which these medicinal plants are harvested from wild habitats [5]. When harvested to an extent, every plant can be exposed to annihilation, although medicinal tree species are most vulnerable to harvesting as they are slow-growing, slow- reproducing and many have specific habitat requirements that limit their distribution [6]. Thus, plants withering owing to harvesting are, therefore, not readily replaced. For centuries the sustainable use of medicinal plants was facilitated by several indirect control methods and some intentional management practices. Some of these practices became unused as development and alteration in traditional healing practices were experienced.

Plant cell culture is a technique that investigates the conditions that influence cell division and genetic regeneration *in vitro* propagation, and it is considered an essential tool in both applied and fundamental studies, additionally as in commercial application [1, 2]. Presently, the facilities for *in vitro* cell cultures are found to be applicable in each plant biology laboratory as a helpful tool for various purposes since tissue culture has turned into a fundamental asset for modern biotechnology, from the critical biochemical aspects to the massive propagation of selected individuals. There are five main areas where *in vitro* cell cultures are being currently applied and can be recognised: As a model system for essential plant cell physiology aspects, generation of genetic modified fertile individuals, large-scale propagation of elite materials, preservation of imperilled plant species, and metabolic engineering of fine chemicals [7].

In addition, *in vitro* culture is a technique that involves the replication of new cells, tissues, and organs derived wholly from the mitotic cell division, consequently generating cloned cells, tissues, and individuals, viz., with identical genetics to the mother plant. Application of *in vitro* culture techniques to effectively produce secondary metabolites, basically plant-derived medicinal compounds, has its main advantages as follows:

1. Decrease in environment interference due to conditions controlled in the *in vitro* culturing chamber.

2. The contingency of having greater control over PDMCs production. Whereby a step-by-step protocol was formed to speed up fresh biomass, accumulation and increase PDMCs concentration in the tissues,

3. The season-independent staggered production of plants-derived medicinal compounds (PDMCs) produces plant-derived medicinal compounds under sterile conditions with fewer risks of contamination by undesired toxic compounds [8].

1.1 Importance of plant tissue culture techniques

Plant tissue culture techniques can be widely used as biotechnological tools for basic and applied purposes. Thus, ranging from investigating plant developmental processes, functional gene studies, commercial plant micro-propagation, and the generation of transgenic plants with explicit industrial and agronomical traits. Then, plant breeding and crop improvement, virus elimination from infected materials to render high-quality, healthy plant material, preservation, and conservation of germplasm of vegetatively propagated plant crops. And save plant species that are endangered by external factors, such as the environment. Additionally, plant cell and organ cultures are interested in producing secondary metabolites of industrial and pharmaceutical interest. Modern technologies, such as genome-editing ones combined with tissue culture and Agrobacterium tumefaciens infection, are currently promising alternatives for the precise genetic manipulation of attractive agronomical or industrial traits in crop plants [7].

1.2 Basic principles of plant cell culture

There are some basic principles of plant cell culture, which are as follows: (1) Select an appropriate explant from a healthy and vigorous plant, (2) eradicate microbial infection of an explant from the surface, (3) inoculate the explant in an adequate culture medium, and (4) provide the explant in culture with the suitable controlled ecological conditions. In the case of *in vitro* regenerated plants, they are subjected to acclimatisation process in the greenhouse before the transference to *ex vitro* conditions. *In vitro* clonal propagation is one of the most current extended commercial applications of tissue culture [9]. These principles are contingent on the part of the plant that is cultured; we can refer to them as cell culture (gametic cells, cell suspension, and protoplast culture), tissue culture (callus and differentiated tissues), and organ culture (any organ, such as zygotic embryos, roots, shoots, and anthers, among others). Each type of culture is used for different basic biotechnological applications [9].

While demand for a consecutive increase in the supply of medicinal herbs to accelerate the evacuation of natural and artificial resources, enhancing various medicinal plants in domestication, adaptation, and cultivation can emerge as an essential strategy for facing the increasing request. Overall, the tendency is towards a more significant proportion of cultivated material in all countries [10]. Most of the national and international companies worldwide, such as over-the-counter markets, the mass and niche markets, and many herb companies, choose cultivated plants because cultivated material could be confirmed as 'organic' or 'biodynamic' [11].

1.3 Practical significance of medicinal plants

In the advanced world of medicine, medicinal plants are essential as raw materials for essential drugs, although synthetic drugs and antibiotics brought about a revolution in controlling different diseases. On the contrary, these synthetic drugs are not accessible to millions of individuals, and some research has shown that thousands of plants contain potent antioxidant compounds, especially phytochemicals and vitamins, as a result of the redox properties they possess and the effect they have to quench singlet oxygen reactive species and tendency to chelate metals [12, 13]. Those living in remote places depend on traditional healers they know and trust. The judicious utilisation of medicinal herbs can even cure deadly diseases that have long defied synthetic drugs evaluated via *in vitro* assays and by *in vivo* supplementation of human and animal models [14]. Some medicinal plants predominately have similar characteristics and components. The likes of Chicory (C. intybus), Stevia (*S. rebaudiana*

Bertoni), Bee Balm/Cambridge Scarlet (Monarda cititrodora), Rhodiola krylovii, and Hedysarum coronarium are found to have some standard bioactive components (see **Table 1**).

*1.3.1 Chicory (*C. intybus*)*

Medicinal plants are predominantly used as food supplements. Chicory (C. intybus) is from the Asteraceae family, a biennial/perennial herbaceous plant, and the stems and leaves are usually eaten as salads. At the same time, the roots are taken as quasi coffee after roasting since there is a similarity between the taste of Chicory to coffee taste, and it is free from caffeine [15]. In medicinal applications, the whole sectors of the Chicory plant have been used, owing to its vital bioactive compounds, such as chicoric acid, vitamins, flavonoids, phenols, sesquiterpene lactones, and fructose polymer inulin, which can act as a sweetener and pre-biotic ingredient [16, 17].

Previous studies have proposed that chicory inulin enriched with oligo-fructose improves calcium absorption and promotes bone maker in the intestines of healthy post-menopausal women [18]. Supplements of chicory help reduce iron overload and aid the proper functioning of the liver [19]. Chicory inulin has a health-enhancing mechanism potential that can disrupt the activity of gut microbiota when it is enriched, and there is a change in its composition [20]. Research has shown that increased intestinal gram-negative bacteria load in persons with diabetes can be linked to the issue of higher lipopolysaccharide (LPS) production, a structural compound in gram-negative bacteria [21]. Hence, data obtained by Landmann et al [22] indicate that chicoric acid can decrease acute alcohol-induced steatosis in mice via induction of iNOS and iNOS-dependent signalling cascades in the liver when it is altered.

*1.3.2 Stevia (*S. rebaudiana *Bertoni)*

The stevia plant has a variety of species that are rich in taste; *S. rebaudiana* is a sweetener plant that is from the genus family Asteraceae and is referred to sweet leaf, sweet weed, honey leaf, or sweet herb. It is the sweetest when compared to other

S/No	Names of Medicinal Plants Used	Characteristics	Components
1	Chicory (*Cichorium intybus*)	Asteraceae family Biennial/perennial herbaceous plant	chicoric acid, vitamins, flavonoids, phenols, sesquiterpene lactones, and fructose polymer inulin
2	Stevia (*Stevia rebaudiana Bertoni)*	Asteraceae family herbaceous plant	Stevioside (glycoside), flavanol, phenols
3	Bee Balm/Cambridge Scarlet (*Monarda citriodora*)	Herbaceous perennial/ biennial plant	Linalool, essential oils, phenols, monoterpene
4	Rhodiola krylovii (Hong Jing Tian)	Herbaceous plant	Salidroside (glucoside), tyrosol, phenols, flavanol, essential oil, monoterpene
5	*Hedysarum coronarium* L.	Herbaceous perennial/ biennial plant	Xanthones, Norisoprenoids, Flavanols, Catechins

Table 1.
Common characteristics of some medicinal plants.

Stevia species. Nevertheless, it has been identified around the globe to have a sweeter glycoside called stevioside that was the primary isolate from it; there are different other sweetness related phytochemicals, such as stevioside, rebaudioside A, B, C, D, E, and glucoside A, which has also been isolated from the leaves of *S. rebaudiana* [23, 24].

*1.3.3 Bee balm/*Cambridge scarlet *(*Monarda citriodora*)*

M. citriodora or *Monarda didyma*, also known as bee balm or Cambridge scarlet, is a herbaceous plant from the subgenus family Cheilyctis and from Aristatae, which originated from North America and was naturalised in Europe. These plant species have been cultivated as garden plants, food, and medicine [25].

Essential oil is a vital constituent of this plant and has been studied by many researchers due to individual species of Monarda [26].

The essential oils obtained from the leaves and flowers of *M. didyma* L. and *M. citriodora* L. cultivated in France compared to others grown in different countries had the same quantitative differences. Thus, the data collected by Collins et al. [27] show a significant component known as linalool, which was found in both flowers and leaves ranging from 64.5% to 74.2%, respectively. In the flowers, y- terpinene was 5.3% and the leaves had 0.9%. The levels of P-cymene in the flowers were 11.0%, which was higher than the composition of p-cymene in leaves at 2.1% [28].

1.3.4 Rhodiola krylovii

Rhodiola is a genus plant called Hong Jing Tian; Crassulaceae comprises over 200 species. Thus, the approximated number of species is 20, which include: *Rhodiola krylovii, Rhodiola rosea, Rhodiola alterna, Rhodiola crenulata, Rhodiola quadrifida, Rhodiola sachalinensis, Rhodiola sacra, and Rhodiola brevipetiolata* [29]. These medicinal plants are cultivated in the Himalayan belt, Tibet, China, and Mongolia. Nonetheless, they are also cultivated in Europe and North America, and these plant products are sold in the market as dietary supplements [30, 31].

Rhodiola plant varies in species, and studies carried out on varieties of Rhodiola have shown salidroside to be available in all the species of the Rhodiola genus. At the same time, rosavin (rosavin, rosin, rosarian) are compounds with a certain amount of *R. rosea* L. [32]. Rhodiola plant, particularly *R. rosea*, is often grown and used in Eastern Europe and Asia to promote physical and mental health. It is cast-off as a traditional medicine for nervous system stimulation to cure depression and fatigue, improve work output, and avoid high-altitude illness, mountain malhypoxia, and anoxia [33]. In contrast, it is used in Russia and Mongolia to treat chronic illness and weakness resulting from pathogenic infection [34]. *R. rosea* has been proven to have cardiovascular protection effects [35, 36]. In addition, the Rhodiola capsule displays anti-depressive potency in patients with depression when administered in dosages of either 0.3 or 0.6 g/day over 12 weeks. Rhodiola capsule tends to improve the eminence of life and clinical symptoms. The high doses of Rhodiola capsules are better than the lower doses [37]. Extensive efforts have been put to cultivate this plant [38].

1.3.5 H. coronarium

H. coronarium L is a perennial forage legume plant usually called French honey suckle, sulla, or sulla clover, and it is a bushy, herbaceous perennial or biennial that typically grows to 3'feet tall with a short life span. The plant is native to Northern

Africa (Algeria (N.), Morocco, Tunisia) and Europe Southwestern Europe, such as Spain. It is a plant that has been cultured as a domestic plant in the 18th century in Southern Italy and as a biennial crop for hay making, grazing, and ensiling [39]. The genus *Hedysarum L.* is approximately made up of 100 species. These are widely spread from the temperate region to boreal regions of the Northern Hemisphere. The plants of this genus occur in numerous habitats, such as deserts or seashores, alpine and arctic meadows, and stony grasslands [40, 41].

Medicinal plants can be protected through increased regulation, and the introduction of sustainable wild harvesting methods, a more viable long-term substitute is to increase domestic cultivation of medicinal plants. Cultivation also opens the possibility of using biotechnology to solve problems inherent in the production of herbal medicines. These include species misidentification, genetic and phenolic variability, variability and instability of extracts, toxic components, and contaminants. Cultivation offers the opportunity to optimise yield and achieve a uniform, high-quality product. However, the prospective cultivator of medicinal plants must make the difficult decision of which species to grow in what is a rapidly shifting and fashion-prone market [42]. Plant-specialised metabolites, also known as secondary metabolites in opposition to so-called primary ones, represent a massive reserve of bioactive compounds amenable for many human applications. Among them, polyphenols are particularly desirable in food crops due to their numerous health benefits, notably their antioxidant properties [43].

The World Health Organisation has valued that over 80% of the world's population in developing countries depend primarily on herbal medicine for basic healthcare needs [38]. However, the use of herbal medicines in developed countries is also growing, and 25% of the UK population takes herbal medicines regularly. Approximately two-thirds of the 50,000 different medicinal plant species in use are collected from the wild, and, in Europe, only 10% of medicinal species used commercially are cultivated [38]. There is concern about decreasing numbers, loss of genetic diversity, local extinctions, and habitat degradation. Well-known species threatened by wild harvesting include *Arctostaphylos uva-ursi* (bearberry), *Piper methysticum*, and *Glycyrrhiza glabra* (liquorice). Thus, between 4000 and 10,000 medicinal species might now be endangered [38, 44].

The main bioactive compounds in medicinal and aromatic plants are secondary metabolites, abiotic stress such as water deficit stress, which tends to have a more significant effect on the medicinal plants' secondary metabolites, and biologically active substances [45]. Phenolic compounds are the essential constituents in the cell defence system against free radicals in abiotic and biotic stresses and are involved in various plant processes, such as growth and reproduction [46]. Results of many studies [45–47] have shown the influences of reduced irrigation or water deficit stress on active substances of medicinal and aromatic crops.

The *in vitro* culturing of medicinal plants depends on the explant and the interaction of the medium. Hence, agar, as a conventional gelling agent, has been reported to have several drawbacks that negatively affect culture growth and differentiation. The gradual uptake of nutrients in the solid medium may lead to lower nutrient availability to the plants. Hence, a reduction in growth rate [48, 49] reported that an agar-solidified medium has lower water availability and uptake by the plants than a liquid medium. This lower uptake of nutrients could explain the lower rate of development of plantlets in solid media compared to liquid media [50]. However, the use of phytagel has not been widely reported. The liquid medium was discovered to cause more roots, nodes, and leaves in the plantlets to sprout than the solid medium; in addition, the liquid medium is cheaper than the solid medium and more economical to use than the solid medium for potato *in vitro* micro-propagation [51].

On the contrary, the factors that affect the number of polyphenols in plant tissues depend on the age or genetic traits of the plant and many external factors, such as microorganism and pest infestation, environmental factors (temperature, humidity, and moisture), which depend on altitude and time of harvest [52–54].

1.4 Composition of culture media for growing medicinal plants

In vitro micro-propagation comes along with plantlet acclimatisation and growth in the greenhouse or the field, which is a prospect that can be incorporated in the product of secondary plant metabolites, especially in rare or endangered species, and in those difficult to propagate [55] and slow to grow [56]. The culture media is made up of minerals; micro (Mn, Zn, Cu, B, Fe, and Mo) and macronutrients (P, K, H, Mg, S, N, and Ca), and vitamins (B1 (thiamin), B6 (nicotinic acid pyridoxine)), growth hormones/regulators/stimulators (auxins, cytokinin, gibberellins, and abscisic acid), and agar (Bacto and Purified) for solid media.

1.4.1 Solid media

Sucrose 3 g, solution of iron chelate 0.5%, macronutrients 5 ml, micronutrients 0.1 ml, vitamins 0.1 ml, Kinetin 0.1 ml (c = 1 mg/ml), 1-naphthalene acetic acid (NAA) 0.4 mg/ml or 0.25 ml, agar 0.9 g, distilled water 100 ml according to Murashige and Skoog, 1962 method for media composition of 100 ml.

1.4.2 Liquid media

Sucrose 3 g, solution of iron chelate 0.5%, macronutrients 5 ml, micronutrients 0.1 ml, vitamins 0.1 ml, Kinetin 0.1 ml (c = 1 mg/ml), 1-naphthalene acetic acid (NAA) 0.1 ml, dichlorophenoxyacetic acid 0.1 ml, distilled water 100 ml according to Murashige and Skoog, 1962 method for media composition of 100 ml.

Furthermore, to find high-frequency adventitious shoot regeneration for similar genotypes, adequate concentrations, and a combination of growth regulators, such as auxins and cytokinin (Kinetin, zeatin, and thidizuron-N-Phenyl, N-1,2,3 thiadiazol-5 urea), should be controlled in this regard. Since the type of plant tissue and concentrations of plant growth regulators in plant cell culture can meaningfully affect the growth morphogenesis of plants [57] Liquid culture system is a critical step to enhance the multiplications rates of shoots produced *in vitro* [58]. The main properties of cytokinin include releasing lateral bud dormancy and stimulating cell division [59].

1.5 Vital components of medicinal plants

Plant growth regulators, including zeatin and thidiazuron (TDZ), and physical, chemical, and biological factors can affect the morphogenesis or organogenesis of plants. Shoot regeneration and development will vary among lingonberry clones [60]. Although most medicinal products are conventional, including those containing molecules derived from medicinal plants, in this case, isolated from the whole, contain a single PDMC as a chemical marker of reference. Cinnamon is an illustrative example of these two treatments: While the use of the cinnamon bark as an infusion to treat infectious diseases characterises the use of the medicinal plant; its primary, secondary metabolite, cinnamaldehyde, which is isolated from the bark, proved to be an efficient antimicrobial agent [61] and can be used as a conventional medicament, similar to other types of synthetic antimicrobials.

*1.5.1 Chicory (*Cichicorium intybus L.*)*

Its species and use categorise Chicory; thus, industrial Chicory, also known as *C. intybus* L. var. sativum, is among the family of the Asteraceae, which is extensively used to produce inulin in South Africa, northern Europe, India, and Chile (Street et al., 2013). Although, the comprehension of the bioactive compounds that undergo synthesis by the biochemical pathways is farfetched [62]. Chicory as an essential plant contains four primary polyphenols: Caftaric, Chlorogenic, Isochlorogenic, and chicoric acids that are prominent in the type of *C. intybus* found in Nord-Pas-de-Calais, France [63]. Caffeic esters in Chicory have been portrayed to have antioxidant properties and potential therapeutic properties, such as anti-diabetic properties [64–66]. One of the essential phenolic compounds in Chicory is chicoric acid, also known as diacetyl tartaric acid; it is used to treat AIDS; thus, it serves as an anti-AIDS agent [67]. The root pulps of Chicory constitute a significant by-product of inulin producing industries and are used as feeds for animals. Extracts from chicory pulps contain a high quantity of pectin, a polysaccharide widely used as a gelling agent, stabiliser, and thickening agent in food [68]. Inulin is a soluble fibre that develops naturally in the chicory plant and has powerful medicinal benefits for human health, it controls and lowers fat, sugar, and calorie in the body, thus giving a tasty appeal, and it can be described as a natural fructan that tends to provide nutritional and health benefits when modified with oligo-fructose than when it is in a pure form [69].

1.5.2 S. rebaudiana Bertoni

S. rebaudiana is referred to as a medicinal plant owing to its natural sweet attribute. This natural resource is recommended for millions of diabetic patients as part of their daily intake since it serves as a natural sweetener and contains substances that promote wellness. Also, the leaves of Stevia comprise flavonoids, antioxidants, alkaloids, water-soluble chlorophylls, xanthophyll, water-soluble inert oligosaccharides, free sugars, amino acids, essential oils, trace elements, vitamins, hydroxycinnamic acids (Caffeic), and polyphenols. Thus, the low-calorie diterpenoid steviol glycosides found in the leaves of Stevia give a sweet taste that is almost 300 times sweeter than usual sucrose [70–72]. Steviol glycosides from the leaves of Stevia are rapidly being developed into an essential ingredient for the food industries for use as sweetener and flavour garnish. The biochemical constituents contained in the plant are of benefit to the pharmaceutical industry [73].

1.5.3 M. citriodora *(bee balm/*Cambridge scarlet*)*

Monarda plant contains an essential oil rich in phenolic monoterpenes, which differs according to its taxon and region of cultivation, and this was shown for medicinal plants. Thus, numerous species, such as *Monard didyma L.* (*M. citriode)* and *Monard fifistulosa* L., are used as medicinal, flavouring, and ornamental plants due to their composition. The hybrid crossing between these species may lead to dynamic hybrids with elevated decorative value and high essential oil contents [74–76]. The Monarda leaves and flowering stems contain water infusions that possess diuretic, anti-helminthic, carminative, expectorant febrifuge, stimulant, and rubefacient properties, which help in the treatment of colds, headaches, and reduce insomnia, also acts as a stomach agent. The plant also possesses strong antifungal activity [77].

1.5.4 Rhodiola krylovii

Rhodiola krylovii is a medicinal plant with several compositions, such as polyphenols, which include; flavonoids, tyrosol, pro-anthocyanidins and cinnamyl alcohol glycosides, organic acids, essential oils, sugars, fats, alcohols, and proteins [78]. Several types of research have demonstrated that the main composition of the Rhodiola plant is tyrosol and salidroside. These compounds possess anticancer bioactivities, antifatigue, antidepressant, antioxidant, adapt genic, anti-inflammatory, and antinociceptive, modulate immune function and prevent cardiovascular, neuronal, liver, and skin disorders [79]. Due to their bioactivity, Rhodiola plant extracts, such as tyrosol and salidroside, tend to stop ageing and attenuate age-related diseases in humans and animals [80].

1.5.5 H. coronarium

H. coronarium (Sulla or French honeysuckle) is typified mainly by high percentages of Norisoprenoids as breakdown products of carotenoids, controlled by vomifoliol. Hence, the other main compounds include 3-hydroxy-4-phenylbutan-2-one and methyl syringate. These compounds are extractable natural volatiles and semi-volatiles distinct compounds, such as a small number of terpenes, norisoprenoids, benzene derivatives, aliphatic compounds, and Maillard reaction products, in the extracts signifies that *H. coronarium* is rather distinctive as compared to the other kinds of honey of the genus that gas chromatography and mass spectro photometer have chemically studied. However, specific markers of the honey botanical origin have not been discovered. In addition, the results obtained show that *H. coronarium* has the potential for a chemical characterisation since reliable data have been obtained so far by using high-performance liquid chromatography and headspace solid-phase micro-extraction in the initial study of *H. coronarium* natural volatiles that is extractable from Sardinian sulla honey samples. Including a variety of distinctive norisoprenoids, benzene derivatives, aliphatic compounds, and Maillard reaction products, but only a few terpenes were found [81, 82].

1.6 Uses and benefits of medicinal plants for medicine

Previous studies by researchers on medicinal plants have claimed that plants have been used for medicine about 60 thousand years ago, which paved the way for more discoveries of medicinal plants [83]. The use of medicinal plants for medication is progressively becoming a primary type of medication, mainly in tropical and underdeveloped countries. Medicinal plant utilisation has been adopted to prevent infectious diseases and avoid death in humans. Whereas, inadequate application of these plants may cause resistance to bacteria, resulting in the antibiotic crisis, along with the limited development of novel molecules [84].

The medicinal applications have been conducted with aid of some medicinal plants. For example, stevia extracts were used as intravenous infusions in rats to reveal their effect on glucose metabolism, diuresis, organ weights, and endocrine function anti-androgenic activity. Stevia extracts have shown some health benefits since they serve as antioxidants and blood pressure and hypertension reducers [85]. The Monarda plant acts against pathogenic and microbial spoilage of food and promotes health. Examples of such organisms that this plant attacks include; Escherichia coli, Listeria monocytogenes, Staphylococcus aureus, and others Salmonella typhimurium or Pseudomonas aeruginosa [86, 87].

Medicinal leaf extracts have antibacterial and fungicidal properties, with an inhibitory effect on the growth of such bacteria as S. aureus, Streptococcus mutants, Bacillus subtilis, and E. coli [88]. Stevia's antihypertensive, anti-inflammatory, anticancer, antidiarrheal, diuretic, and immunomodulatory properties have been shown to this extent [70]. Rhodiola plant extracts are utilised traditionally in tonics and adaptogens and are incorporated in antidepressant and anti-inflammatory drugs [89, 90].

1.7 Factors influencing the cultivation of medicinal plants

Some factors influence the growth of plants (medicinal plants), which may affect many features and the quality of these plants. These factors include temperature, light, altitude, atmospheric humidity, soil, and rainfall:

a. Temperature: It is one of the significant factors that influence the cultivation of medicinal plants. An increase and decrease in temperature affect photosynthesis, and plant respiration rate is affected by causing an increase in respiration rate.

b. Light: It is an essential factor for medicinal plant cultivation growth. It aids the process of photosynthesis, germination process of seeds, flowering, vegetative growth, tuber formation, and the opening and closing of stomata in plants—the growth rate of plants increases during the night since the dark phase. Hence, High light intensity affects plant growth.

c. Altitude: It is a vital factor that influences the cultivation of medicinal plants. As a result of an increase in altitude, there is a decrease in temperature and atmospheric pressure. While the velocity of wind, relative humidity, and light intensity increase. Vegetative properties can change as a result of changes in altitude.

d. Atmospheric Humidity: It is an essential factor that can influence the transpiration rate in plants, thereby causing water evaporation from the earth's surface.

e. Soil/Nutrient Composition: Plants are grown on soil or in a substrate that contains vital nutrients. Thus, selected medicinal plant species species-specific medicinal plant parts can dictate optimal soil conditions, including soil type, drainage, moisture retention, fertility, and pH. The addition of excess fertiliser can accumulate soluble salts, which form a whitish crust on the growing medium's surface.

f. Water/Rainfall: In an external setting, rainfall influences plants' growth, such as the morphology and physiology of plants. For instance, continuous rainfall can cause water-soluble substances from the leaves and roots by leaching to be lost, mainly associated with plants that produce glycoside and alkaloids [91]. High production and labour costs limit the commercial use of micropropagation [92].

1.8 Limitations of medicinal plant cultivation

There are shortcomings in medicinal plants' applications due to their exceptional qualities in the class of plants, and the products from these medicinal plants are essential sources for improving and maintaining human health [93]. Medicinal plant cultivation has been abortive due to the complexity of germination and environmental factors [38].

It can be clearly stated that the information about the use of medicinal plants is enormous, and in some circumstances, these natural resources are incorporated as the only form of preventing and curing diseases [94]. Anxiety around the introduction of medicinal plants in health systems is about how people utilise these products and how they will be recommended and presented to the population. Since specific consumers who may have some empirical knowledge about the use of certain medicinal plants may fall short of information concerning the actual toxicity of several species, a large number of the population still think that plants cannot damage their health. However, not all the users of these plants have this ideology about these natural resources.

What is more, is the fact that medicinal plants have a variety of species that are known by the same name and when misused, can lead to allergy and stimulation of medical conditions. There is inadequate knowledge of how the preparation and contraindications of the plants, lack of knowledge by health professionals regarding the side effects after consumption, difficulty regarding correct usage of these natural resources, and unreliable information about these plants, which leads to difficulty disseminating this to people. Complementary practices in administering these natural resources to patients to treat specific ailments are abandoned [95].

The storability of medicinal plants plays a role in the ineffective use of the products since many biochemical differences can occur within the natural resources leading to changes in bioactivity [96].

2. Materials and methods

2.1 Experimental location

Experiments were conducted in the Department of Technology for Organic Synthesis, Institute of Chemical Engineering Ural Federal University, Ekaterinburg, Russia.

2.2 Raw material location

The raw materials included *C. intybus, S. rebaudiana Bertoni, Monarda citriodora, Rhodiola krylovii Chicory, Stevia, Cambridge scarlet (M. didyma), Rhodiola krylovii* and were all obtained within Russian Federation.

2.3 Raw material consideration

The primary raw material consideration for the *in vitro* cultivation of these plant cells includes solid and liquid culture media with enough nutrient composition required for the plant cells' growth.

2.3.1 Materials/equipment for experiment

2.3.1.1 Reagents for media

Sucrose, solution of iron chelate, macronutrients, micronutrients, vitamins, Kinetin, 1-naphthalene acetic acid (NAA), dichlorophenoxyacetic acid, agar, and distilled water 100 ml.

2.3.2 Reagents for plant cells sterilisation

Ethanol (70%), sodium hypochlorite (3%), Tween detergent, sterile water, carboxylic acid amide (CAA) fungicide (2%).

2.3.3 Equipment

Climate chamber, Petri dishes, laminar system, test tubes, autoclave, oven, stirrer, laboratory glassware (conical flask, beakers, burette), micropipette and pipette tips, knife, forceps, filter paper, foil paper, electric stove, refrigerator, pincers, Bunsen burner, thermometer, and pH meter.

2.3.4 Methods

Different methods/procedures were used to cultivate *C. intybus, S. rebaudiana Bertoni, Monarda citriodora and Rhodiola krylovii Stevia, Chicory Cambridge scarlet, and Rhodiola krylovii* to identify which sterilisation method was best for each plant type and which procedure would not incur contamination. Studying the effectiveness of various sterilisation methods for seed cultivation on solid media in tubes for callus formation (medium M&S + kinetin) and cultivating seeds on solid media in tubes to obtain sterile micro plants (medium M&S) were carried out. The germination of non-sterile/sterile seeds and obtaining non-sterile and sterile plant seedlings in the Petri dish were carried out. Growing sterile micro explants for callus formation (medium M&S + kinetin) and preparation of cell suspension cultures for wet and dry weight determination.

2.3.5 Media formulation

The media for this research was formulated from Murashige and Skoog Culture Media (1962). The Murashige and Skoog Medium is a shared medium used to cultivate laboratory plant cells. The discovery was made by two Plant scientists, Toshio Murashige and Folke K. Skoog [97], searching for a new type of plant growth regulator. Using Murashige and Skoog medium supplemented with 3 g of sucrose, 0.5 ml solution of iron chelates, 5 ml of macronutrients, 0.1 ml of micronutrients, 0.1 ml of vitamins, 0. ml of Kinetin, 0.25 ml of 1-naphthalene acetic acid (NAA), 0.8 to 0.9 ml of agar, and 100 ml distilled water for every volume of 100 ml media.

2.3.6 Preparation of solid culture media using Murashige and Skoog method

A total of 100 ml of distilled water was poured into a 250 ml round bottom flask. A weighing balance was used to measure 0.8 g of agar added to the water. The sample was placed on a heating device (infrared cooker) at the temperature of 60°C; 0.25 ml of 1-naphthalene acetic acid (NAA) was measured with a micropipette and poured into the sample, and stirred. Then, 3 g of sucrose was added to the sample on heating and stirred, followed by adding 0.1 ml of Kinetin and 0.5 ml of chelate were added to the heating sample. After 10 mins, the sample was removed from the heating device and allowed to cool at room temperature (25°C). Then, 5 ml of macronutrients, 0.1 ml of micronutrients, and 0.1 ml of vitamin were added to the sample as the temperature reached 40°C. The sample was stirred and covered with cotton and foil paper with a plastic band around the cover and kept in the refrigerator to solidify for over 12 hours. The media was sterilised in an autoclave for 1 hour at 121°C. The sample was placed on a heating device for 20 mins at 60°C, and after heating, the sample was left to cool for 10 mins before transferring to test tubes.

2.3.7 Preparation of liquid culture media using Murashige and Skoog method

A total of 100 ml of distilled water was added to a 250 ml round bottom flask. The sample was placed on a heating device (infrared cooker) at a temperature of 60 °C and 0.125 ml of 1-naphthalene acetic acid (NAA) and 0.1 ml of dichlorophenoxyacetic acid, whereas measured with a micropipette and poured into the sample and stirred. Next, 3 g of sucrose was added to the sample while heating and stirring.

Then, 0.1 ml of Kinetin and 0.5 ml of chelate were added to the heating sample. After 10 mins, the sample was removed from the heating device and allowed to cool at room temperature before adding 5 ml of macronutrients and 0.1 ml of micronutrients. A total of 0.1 ml of the vitamin was added to the sample when the temperature was 40° C. The sample was stirred and covered with cotton and foil paper with a plastic band around the cover before it was preserved and kept in the refrigerator to solidify for more than 12 hours. The media was sterilised in an autoclave for 1 hour at 121°C. The sample was placed on a heating device for 20 mins at 60°C, and the sample was left to cool for 10 mins before transferring into a sterile flask. Different for callus cultivation.

2.3.8 Experimental design

2.3.8.1 Seed sterilisation

Studying the effectiveness of various sterilisation methods (A, B, C, and D) for seed cultivation on solid media in tubes for callus formation (medium M&S + kinetin) and the cultivation of seeds on solid media in tubes to obtain sterile micro plants (medium M&S).

I. Using 0.8 grams of agar in the solid media (the amount of agar used for inoculation of cells was 0.8 grams).

1. **Method A – Sterilisation and Cultivation Process of Stevia, Chicory, Bee Balm (Cambridge Scarlet), and Rhodiola krylovii Cells.**

   The volume of 400 ml of media with Kinetin was prepared using the above reagents for plant cells. However, 7 ml of solid media was poured into 52 test tubes. Then, sterile Petri dishes were used for the sterilisation process of seeds and seeds were first sterilised with NaOCl for 3 mins. Then, the seeds were transferred into Petri dishes containing ethanol for 1 min. The seeds were rinsed with sterile water for 2 mins. The seeds were left to dry before transferring them into the test tubes. Two seeds of Stevia, Chicory,

Bee Balm (Cambridge Scarlet), and Rhodiola krylovii cells were put into 13 labelled separate test tubes for each type of plant and were labelled. The test tubes containing samples were incubated with a light intensity of 16 hours and dark phase of 8 hours, a temperature of 250C (±10C) and relative humidity of 50%. After 14 days, a percentage of contamination was carried out.

2. **Method B – Sterilisation and Cultivation process of Stevia, Chicory, Bee Balm (Cambridge Scarlet), and Rhodiola krylovii Cells.**

   The 600 ml of M and S media with Kinetin was prepared using the above 11 reagents for plant cells. A total of 7 ml of solid media was poured into 80 12 test tubes. At the same time, sterile Psetri dishes were used to sterilise seeds. Seeds were sterilised using 70% ethanol for 2 mins. The rinsing of seeds was done with sterile water for 1 min. Seeds were washed with NaOCl for 1 min. Then, rinsing of seeds with sterile water for 1 min. The seeds were allowed to dry for a few mins. Two seeds were put in each test tube (20) for each type of plant and well labelled. The samples were placed in the climate chamber for incubation. Then, the test tubes containing samples were placed in the climate chamber and incubated under conditions for incubation with 16 hours of light phase and 8 hours of no light phase and temperature 250C (± 1°C) and relative humidity of 50%. After 14 days, the percentage of contamination was determined.

II. Using 0.9 g of agar in the solid media (the amount of agar was increased from 0.8 grams to 0.9 grams to reach average thickness).

1. **Method C – Sterilisation and Cultivation Process of Stevia, Chicory, Bee Balm (Cambridge Scarlet), and Rhodiola krylovii Cells.**

   The 400 ml of M and S media with Kinetin was prepared using the above reagents for plant cell culture. A total of 9 ml of media was poured into 40 sterilised test tubes. Sterile Petri dishes were used for the sterilisation process of seeds. Seeds were thoroughly washed in flowing water. Afterwards, seeds were sterilised with 2% carboxylic acid amide (CAA) fungicide for 2 mins, followed by washing seeds with 70% ethanol for 2 mins, then 3% NaOCl solution for 2 mins. Rinsing seeds thoroughly with sterile water thrice (three times) in three separate sterilised Petri dishes for 5 mins. Two seeds of Chicory and Stevia were put into 10 different test tubes containing solid media, respectively. Then, samples were placed in the climate chamber for incubation with no light intensity, temperature 250C (± 1 °C), and relative humidity of 50%. After 14 days, the percentage of contamination was determined.

2. **Method D – Sterilisation and Cultivation Process of Stevia, Chicory, Bee Balm (Cambridge Scarlet), and Rhodiola krylovii Cells.**

   The 400 ml of M and S media volume with Kinetin was prepared using the above reagents for plant cells. A total of 9 ml of solid media was poured into 56 test tubes. Sterile Petri dishes were used for the sterilisation process of seeds. Seeds were washed adequately with an antibacterial soap in flowing water for 10 mins. Seeds were sterilised with 70% ethanol for 2 mins. Seeds were properly washed three times with sterile water. Then, the seeds were sterilised with 3% NaOCl solution for 5 mins. Seeds were later transferred into sterile test tubes containing solid media. Fourteen labelled, sterile test tubes contained Stevia, Chicory, Bee Balm (Cambridge Scarlet), and *Rhodiola krylovii* cells. Seeds were kept in the climate chamber for incubation with no light intensity, temperature 250C (± 1 ° C), and relative humidity of 50%. After 14 days, the percentage of contamination was determined.

The level of percentage contamination for the four sterilisation methods (A, B, C, and D) for seed cultivation on solid media in tubes for callus formation (medium M&S + kinetin) and the cultivation of seeds on solid media in tubes to obtain sterile micro plants (medium M&S) was determined by using the formula:

$$\% \text{ of Contamination} = (\text{Number of Contaminated Test tubes with cells})/(\text{Total Number of Test tubes with seeds}) * 100 \quad (1)$$

2.3.8.2 The germination of non-sterile/sterile seeds and obtaining non-sterile and sterile plant seedlings in Petri dishes

A. Non-sterile Plant Seedlings

The seeds of Cambridge scarlet, Rhodiola krylovii, and Chicory were put into six Petri dishes. Two Petri dishes for Bee Balm (Cambridge scarlet), two for Rhodiola krylovii, and two Petri dishes for Chicory were prepared. The seeds were unwashed, and 7 ml of distilled water was added to all Petri dishes. The samples were labelled and sealed with paraffin and incubated for 2 weeks for plant growth in a climate chamber for a plant with no light intensity, a temperature of 250C (±1°C), and relative humidity of 50%.

B. Sterile Plant Seedlings

A total of 7 ml of sterile water was poured into six sterile Petri dishes. Seeds of Chicory, Rhodiola krylovii, and Cambridge scarlet were disinfected with 70% ethanol for 2 mins and 3% NaOCl solution for 2 mins. Seeds were rinsed thoroughly with sterile water twice in separate sterilised Petri dishes for 2 mins. The seeds were poured into the six sterilised Petri dishes and sealed with paraffin. Samples were labelled and incubated in the climate chamber for 2 weeks with no light intensity, temperature 25°C (1°C), and relative humidity of 50%.

2.3.8.3 Growing sterile explants for callus formation (Medium M&S + kinetin)

I. Growth Media with Kinetin (+Kinetin)

The media for growing callus was slightly different, with a variance in supplements (Murashige and Skoog media, 1962). A 300 ml M and S media volume was prepared, and 9 ml was poured into 24 sterile test tubes. The non-sterile seedlings from the six Petri dishes containing Chicory, Rhodiola krylovii, and Cambridge scarlet were transferred to the test tubes after cutting and disinfection: First roots of seedlings were cut, and explants were disinfected using 2% carboxylic acid amide (CAA) fungicide for 2 mins, followed by the washing of explants in 70% ethanol for 2 mins. Explants were rinsed with sterile water. Seeds were washed with 3% NaOCl for 2 mins and rinsed with sterile water. Explants were transferred into labelled sterile test tubes and were incubated for 2 weeks with conditions of no light intensity, a temperature of 250C (±1 °C), and relative humidity of 50%.

II.Growth Media without Kinetin (-Kinetin)

The volume of 600 ml of M and S media without Kinetin was prepared using the above reagents for plant cells. Then, 9 ml of media was poured into 60 sterilised test tubes each. Sterile seedlings from the six Petri dishes containing Chicory, Bee balm (Cambridge scarlet), and Rhodiola krylovii were cut and thoroughly washed. Disinfection of explants was done using 2% carboxylic acid amide (CAA) fungicide for 2 mins followed by washing with 70% ethanol for 2 mins and then with 3% NaOCl solution for 2mins. Rinsing of explants thoroughly with sterile water thrice (three times) in three separate sterilised Petri dishes for 5 mins was carefully carried out. The three plants' explants were put into 20 labelled sterilised test tubes, respectively, and the remaining media was preserved in the refrigerator. Incubation of explants was done under conditions of no light intensity, the temperature of 25°C (1°C), and relative humidity of 50% for a period of 2 weeks.@

The formula below was used to determine the percentage contamination for the above experiments:

$$\text{\% of Contamination} = (\text{Number of Contaminated Test tubes with cells}) / (\text{Total Number of Test tubes with cells}) * 100 \quad (2)$$

2.3.8.4 Preparation of cell suspension cultures

The callus of Chicory was transferred into two sterile flasks containing M and S cell suspension cultures of Kinetin and without Kinetin. Samples were crushed with a knife and stirred vigorously to aid the formation of more cells. The samples were incubated in plant-controlled equipment under conditions of continued shaking with 16 hours of light intensity phase at 24(±1)C and 8 hours of dark phase at 18 (±1) 0C and after 1 week, the callus was transferred to a new cell suspension culture maintaining the conditions for a week. One week later, the callus in both flasks was transferred to six sterile flasks using the same media formulation; three flasks had no kinetin, while three had Kinetin in the media. A total of 10 ml of starter culture (cell suspension) and 50 ml of liquid media were poured into the six labelled sterile flasks, and the samples were returned to the plant-controlled equipment to grow to maintain the conditions of continuous shaking with 24 hours of light at 24 °C (±1) for 4 weeks. Each week the samples were checked to ensure that they were free from contamination.

2.4 Microscopy

The overall magnification can be calculated as the product of the lenses and the distance over which the image is projected:

$$M = (D * M1 * M2)/250 \text{ mm} \quad (3)$$

where D = projection (tube) length (usually = 250 mm); M1, M2 = magnification of objective and ocular. Thus, 250 mm = minimum distance of distinct vision for 20/20 eyes [98]. Callus of Chicory grown using cell suspension culture (liquid media) of + Kinetin and – Kinetin was viewed under the microscope using a 10 mm focal point.

2.5 Wet and dry weight analysis

Drying process was achieved by initially setting the oven temperature at 60°C. 5 ml of Kinetin containing media starter was pipetted into three separate centrifuge tubes and properly labelled. Afterwards, 5 ml of media without Kinetin was pipetted into three different and well labelled centrifuge tubes. Centrifugation was done for 10 mins on 5000 rpm at 15°C; thus, the aliquots were separated, and 5 ml of sterile water was added to the supernatants and centrifugation was done twice. Six filter papers (2.5 cm) were dried in the oven for 9 hours at 60°C. Then, the empty weight of filter papers was obtained, followed by the weight of filter papers with samples. The samples were later placed in the oven for 24 hours at 60°C. The weight of the dried samples on the filter papers were obtained. The fresh and dry weights were determined one week after transferring the seeds into a new culture media.

$$\text{Wet wt } (g/L) = (\text{Wt of wet filter paper} + \text{cells } (g) \text{ wt of wet filter paper}(g))/(\text{Sample vol } (mL)) \text{ x } 100 \quad (4)$$

$$\text{Dry wt } (g/L) = (\text{Wt of dry filter paper} + \text{cells } (g) \text{ wt of dry filter paper}(g))/(\text{Sample vol } (mL)) \text{ x } 100 \quad (5)$$

2.6 Charts

Bar charts and area charts were used to present the data obtained from the experiments.

3. Results and discussion

3.1 Seed sterilisation: Studying the effectiveness of various sterilisation methods

Results were obtained for the four sterilisation methods after 14 days of cultivating seeds of Stevia, Bee balm, Chicory, and *Rhodiola krylovii*, and the values were presented using tables and bar charts (see **Table 2**).

3.1.1 Method a

The result of method A as shown in **Figure 1**, Stevia cells had 53.9% contamination, which was the least contaminated. While Bee balm, Chicory, and *Rhodiola krylovii* had

PLAN TS	METHOD A			METHOD B			METHOD C			METHOD D		
	Total No. of Test tubes with micro plant s	No. of Conta minat ed. Test tubes	No. Non-conta minat ed. test tubes with micro plant s	Total No. of Test tubes with micro plant s	No. of Conta minat ed. Test tubes	No. Non-conta minat ed. test tubes with micro plant s	Total No. of Test tubes with micro plant s	No. of Conta minat ed. Test tubes	No. Non-conta minat ed. test tubes with micro plant s	Total No. of Test tubes with micro plants	No. of Conta minat ed. Test tubes	No. Non-conta minat ed. test tubes with micro plant s
Stevia	13	7	6	20	11	9	10	5	5	14	2	12
Bee Balm	13	13	0	20	20	0	10	10	0	14	14	0
Chico ry	13	12	1	20	18	1	10	3	7	14	7	7
Rhodi ola	13	13	0	20	20	0	10	10	0	14	14	0

Table 2.
Effectiveness of various sterilisation methods (A, B, C and D).

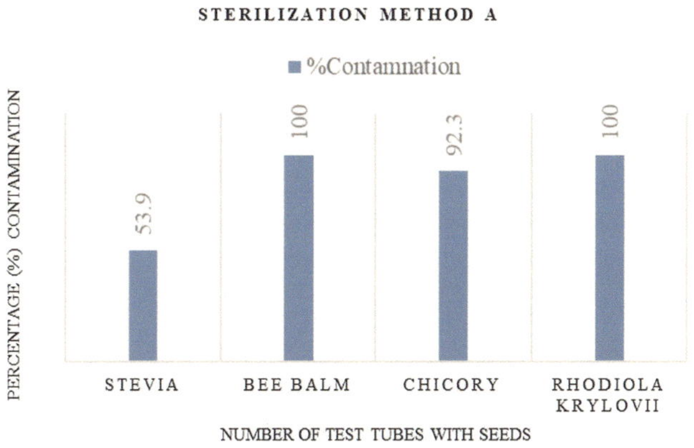

Figure 1.
Chart showing percentage contamination of method A.

100%, 92.3%, and 100%, 92.3%, and 100% contamination, respectively. The result shows that this method is averagely suitable for Stevia seeds but not suitable for Bee balm, Chicory, and Rhodiola krylovii seeds. The sterilisation method is not effective for all the seeds since the level of percentage contamination is high, and the agar composition in this 05 method was 0.8 g.

3.1.2 Method B

The result of sterilisation in Method B **Figure 2** shows that Chicory had 40% contamination and Stevia had 55% contamination. In contrast, Bee balm and *Rhodiola krylovii* had a 100% level of contamination. This sterilisation method of plant seeds is suitable for Chicory (40%) since the contamination percentage was few. Stevia seeds also can be sterilised using this method because the percentage level of contamination after cultivation was 55%. Nevertheless, this method is unsuitable for Bee balm and *Rhodiola krylovii* since the level was very high (100%), as shown in **Figure 2** and **Table 3**, and the agar composition in this method was 0.8 g, and this could have affected the growth rate of the plants.

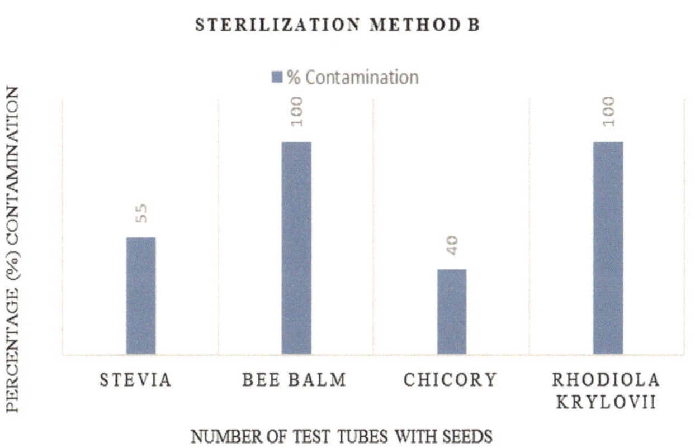

Figure 2.
Chart showing percentage contamination of method B.

PLANTS	Culture Media with Kinetin (+Kinetin)			Culture Media without Kinetin (-Kinetin)		
	Total Number of Test tubes with cells	Number of Contaminated Test tubes	Number of Non-contaminated Test tubes	Total Number of Test tubes with cells	Number of Contaminated Test tubes	Number of Non-contaminated Test tubes
Chicory	8	5	3	20	20	0
Bee Balm	8	6	2	20	20	0
Rhodiola krylovii	8	5	3	20	20	0

Table 3.
Growing sterile explants for callus formation (Solid Medium M&S + kinetin and – Kinetin).

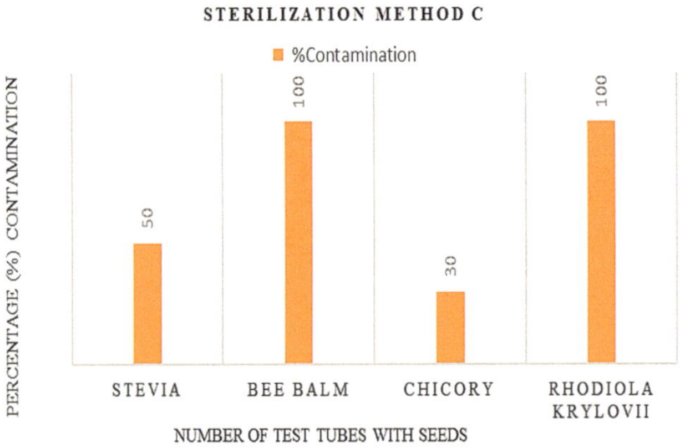

Figure 3.
The chart showing percentage contamination of method C comes after the subheading method C.

3.1.3 Method C

The percentage (%) contamination was determined for method C, and from the result, as shown in **Figure 3**, Chicory and Stevia had 30% and 50% contamination, respectively, while Bee balm and *Rhodiola krylovii* both had 100% contamination. Method C is a suitable sterilisation method for Chicory seeds. The agar composition used for media preparation in this method was 0.9 g, which could have influenced the plant seeds.

3.1.4 Method D

Figure 4 shows the level of percentage (%) of Stevia (14.29), Bee balm (100%), Chicory (50%), and Rhodiola krylovii (100%). The result obtained using method D indicates that Stevia seeds had a minor percentage (%) contamination, which is 14.29%, and it is the best method for sterilising stevia cells. Although Chicory seeds can also be sterilised using this method, it stands an average chance of not getting contaminated than Stevia, which has a greater chance of no contamination. While Bee balm and Rhodiola krylovii had 100% contamination, thus this method of sterilisation is not suitable for both plant seeds, and the agar composition used during the media preparation was 0.9 g.

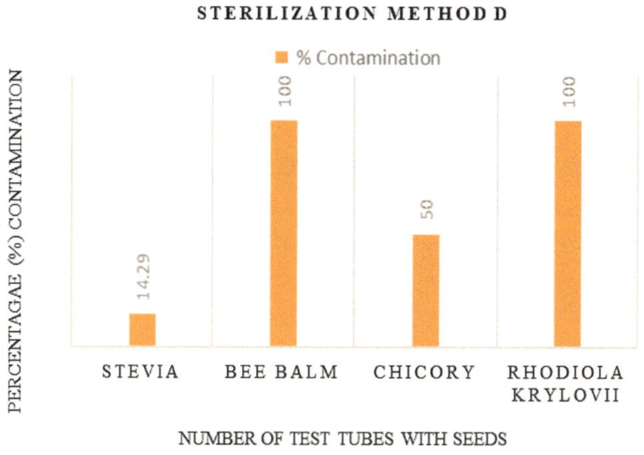

Figure 4.
Chart showing percentage contamination of method D.

From the four (4) methods of sterilisation used in this experiment, the result for each Method (A, B, C, and D) Stevia and Chicory cells were less prone to 100% contamination when compared to Bee balm and Rhodiola krylovii, which turned out to have 100% contamination in the four methods of sterilisation and cultivation. The most suitable method for Stevia is method D which only had 14.29% contamination, and this result can be linked to the findings of Halim et al., 2016 for the sterilisation process of Stevia, which had 15% contaminated plant cell culture. The best method for cultivating Chicory cells was method C, which had a 30% (%) contamination level. Hence, it can be stated that different plant cells have a specific method of sterilisation that enables cells/seeds to be free from pathogenic microorganisms that may affect the cultivation process. However, the composition of culture media might have influenced the growth rate of the plants, and this agrees with the findings of Yildiz and Usha et al. [57, 99] that the higher the composition of plant regulators used, the higher the observable differences and the amount of agar used can affect the growth of plant cells. A total of 0.9 g of agar was used in methods C and D. In comparison, 0.8 g was used for methods A and B. Although, other factors, such as percentage concentration, application period, and temperature of NaOCl could, affect *in vitro* germination of seeds, regeneration potential of explants, and growth of seedlings when these factors are not taken into consideration. Hence, these results might also lead to seed contamination due to the inefficacy of NaOCl used during seed disinfection [100] (see **Figures 5–8**).

3.2 Germination of non-sterile/sterile seeds and obtaining non-sterile and sterile plant seedlings in Petri dishes

Two factors were considered under this experiment; a) non-sterile plant seedlings and b) sterile plant seedlings, and the plants were observed after 2 weeks of culturing. The observable differences between the plant cells of Chicory, Bee balm, and *Rhodiola krylovii* were determined by sight. The variation of growth between the plant cells was evident to the eyes. This experiment was done to obtain callus. Non-sterile Chicory showed moderate growth in the Petri dishes, while the growth of the pure Chicory was prominent. The non-sterile Bee balm showed no growth in the petri dish, unlike the sterile Bee balm was showed noticeable growth (sprouts). Non-sterile *Rhodiola krylovii* showed noticeable sprouts, while sterile *Rhodiola krylovii* showed noticeable sprouting.

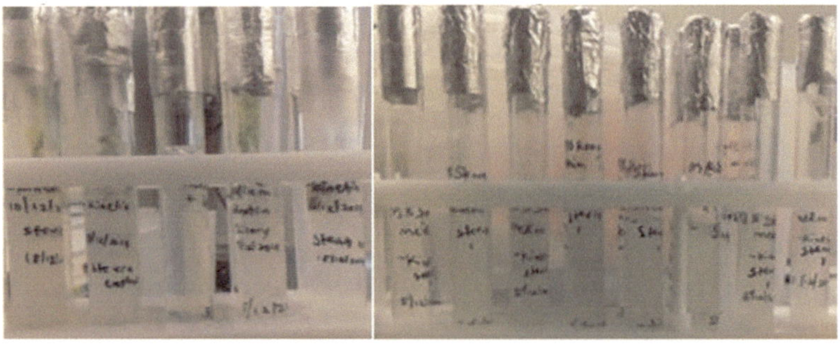

Figure 5.
Cultured stevia seeds in solid media.

i) ii) iii)

Figure 6.
(i) Stevia, (ii) chicory, and (iii) Hedysarum.

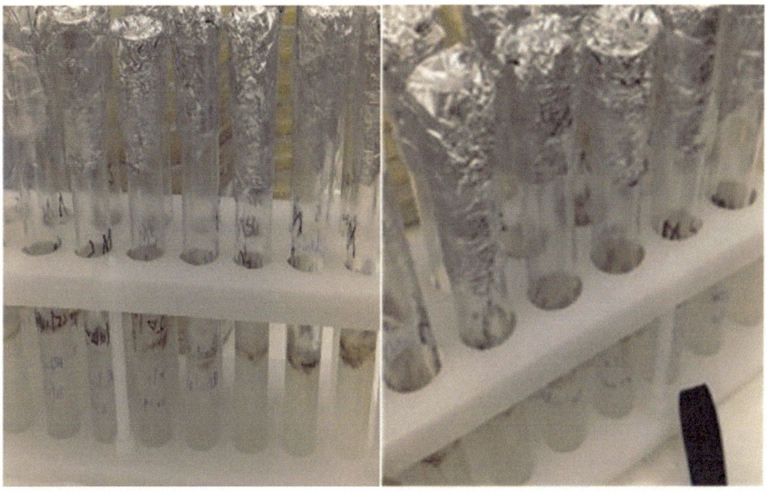

Figure 7.
*Contaminated samples of bee balm (*Cambridge scarlet*) and* Rhodiola krylovii *plants.*

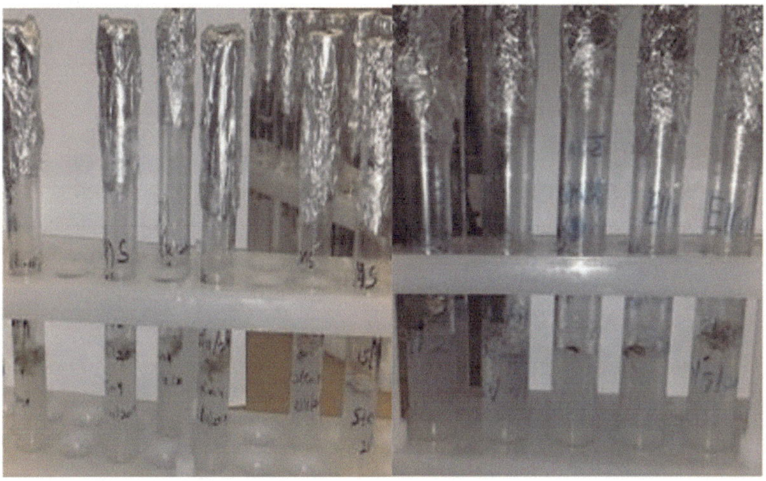

Figure 8.
(i) Growing stage of stevia and (ii) growing stage of chicory.

Hence, this indicates that sterilisation of plant cells is essential in the *in vitro* process of culturing seeds, and basic steps must be considered during sterilisation. The difference in growth between the sterile and non-sterile plant cells might result from their genetic differences among the two subspecies of the used plants. This result agrees with the findings of Debnath and McRae [60], which explained that different genotypes variably react for shoot proliferation and regeneration in a liquid medium or a semi-solid media in his work on a semi-solid medium between subspecies. Also, the water absorption rate can influence the seedlings' rate during *in vitro* cultivation of plant cells [101].

3.2.1 Non-sterile seedlings

The seeds were not sterilised (see **Figure 9**).

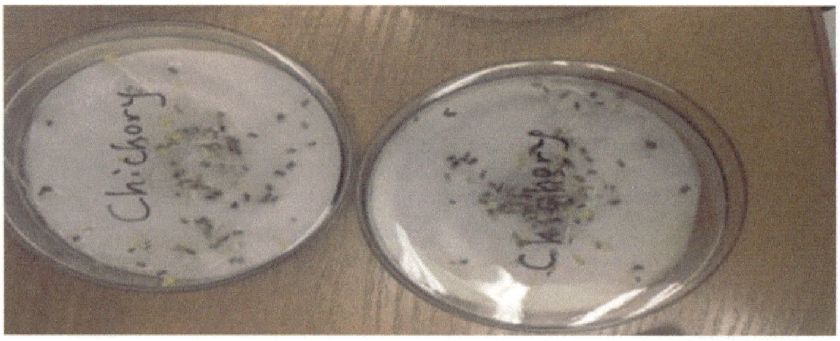

(i)

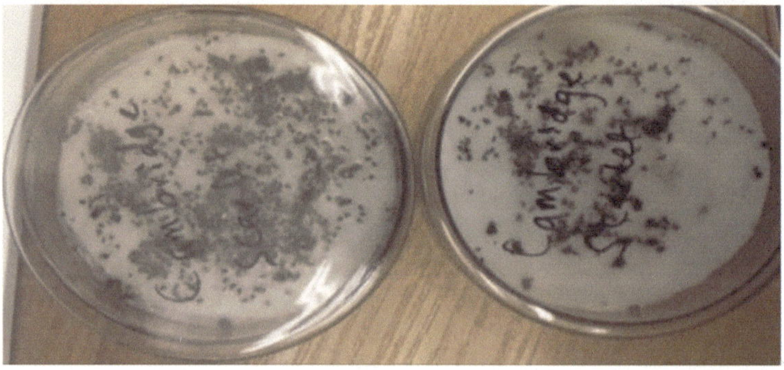

(ii)

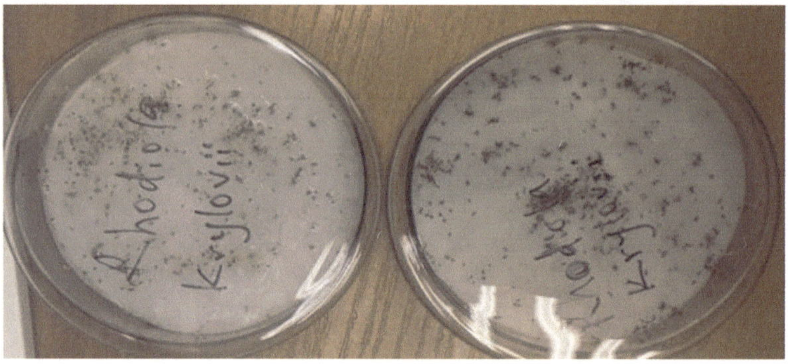

(iii)

Figure 9.
(i) Non-sterile Chicory with sprouts after 2-weeks. (ii) Non-sterile Cambridge scarlet without sprouts after 2-weeks. (iii) Non-sterile Rhodiola krylovii with sprouts after 2-weeks.

3.2.2 Sterile plant seedlings (seeds were sterilised)

No Contamination after 2 weeks (see **Figure 10**).

3.3 Growing sterile explants for callus formation (Solid Medium M&S + kinetin and - kinetin)

Two factors were considered during the cultivation process of Callus of Chicory, Rhodiola krylovii, and Beebalm. These factors were (I) Growth Media with Kinetin

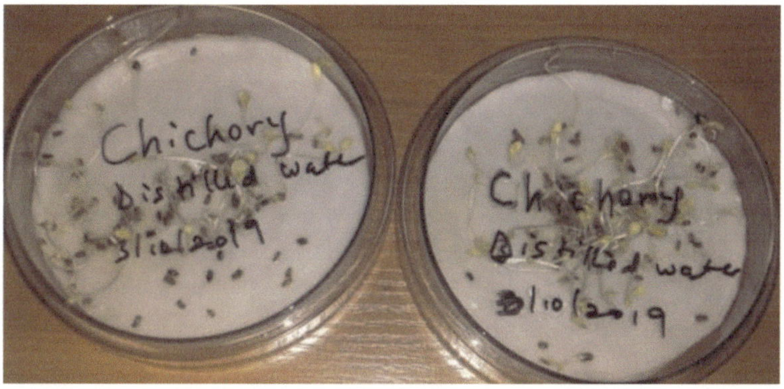

(i)

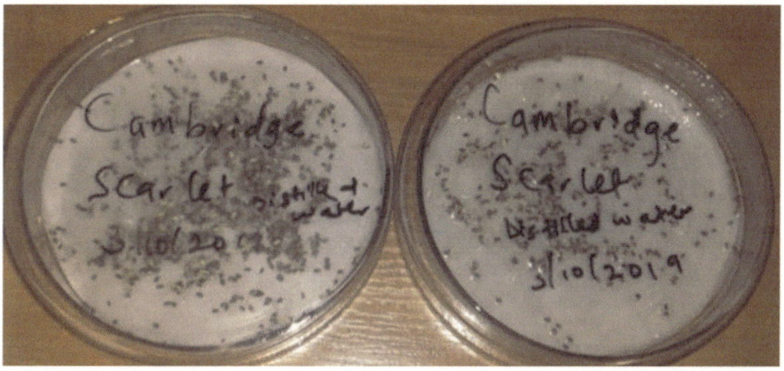

(ii)

(iii)

Figure 10.
(i) Sterile Chicory with sprouts after 2-weeks. (ii) Sterile Cambridge scarlet with sprouts after 2-weeks. (iii) Sterile Rhodiola krylovii with sprouts after 2-weeks.

(+ Kinetin) and (II) Growth Media without Kinetin (-Kinetin) and the determination of percentage (%)contamination of callus in test tubes after 2 weeks was done to know the percentage contamination of each plants using different media.

Growth of callus was observed with samples in media containing Kinetin. However, there was 62.5–75% contamination in the media containing Kinetin, as shown in **Figure 11**. Chicory and Rhodiola krylovii had the same level of percentage contamination, which was 62.5% and less than the percentage contamination of Bee balm.

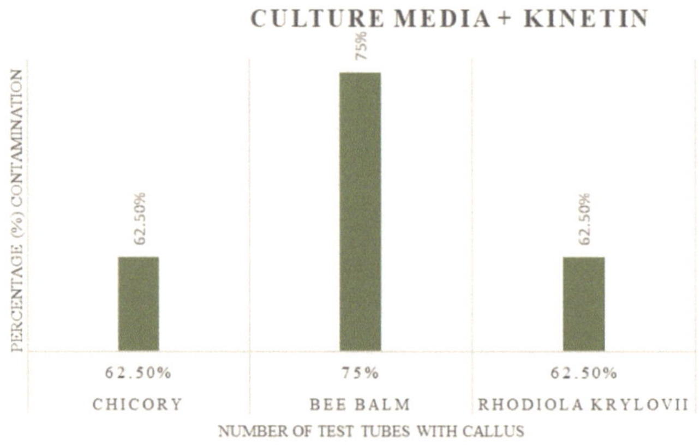

Figure 11.
Percentage contamination for callus formation (Medium M&S + kinetin).

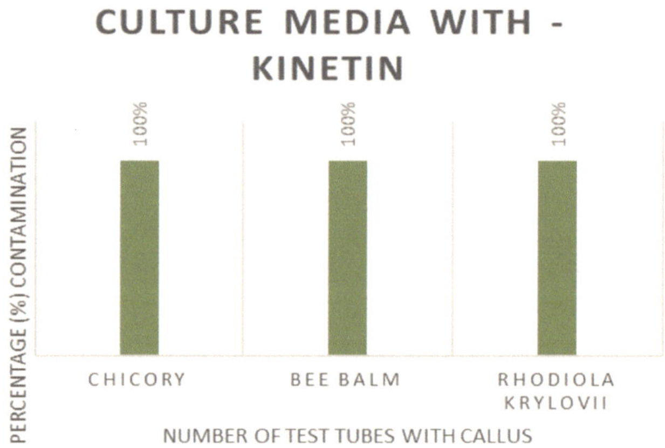

Figure 12.
Percentage contamination for callus formation (Medium M&S - kinetin).

After 2-weeks, no sign of growth was observed from any of the samples in the media with no Kinetin and all samples were contaminated. As shown in **Figure 12** of growth media without Kinetin, the level of contamination was 100% for all plants, and no growth was observed in all plant cultures. Thus, this can be due to the absence of Kinetin, a plant hormone that helps induce callus and shoot regeneration formation. While the high level of contamination can be a result of the poor method of sterilisation and the efficacy of sodium hypochlorite, this agrees with the studies of Racoppi [102], which reported that the sterilising agent NaOCl (Sodium Hypochlorite) could decrease in its activity when all the parameters, such as temperature and pH, are not stable. However, if the temperature is slightly above 10 °C, this might increase disinfection activity since this enables easy incursion into the seed coat. Higher temperatures of NaOCl affected the morphology of seedlings, resulting in abnormal growth of hypocotyls and other parts of the plant cells. The unsuccessful result of no growth must have been due to variation in room temperature. According to Yildiz [57], change in temperature of (±1°C) can have an adverse effect on the plant, resulting in no growth (see **Figures 13–15**).

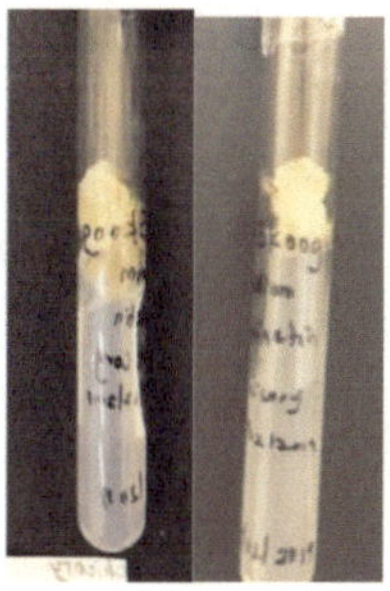

Figure 13.
Callus of chicory in solid culture media after 8 weeks.

Figure 14.
*Bee balm (*Cambridge scarlet*) and* Rhodiola krylovii *in solid media with no growth of callus.*

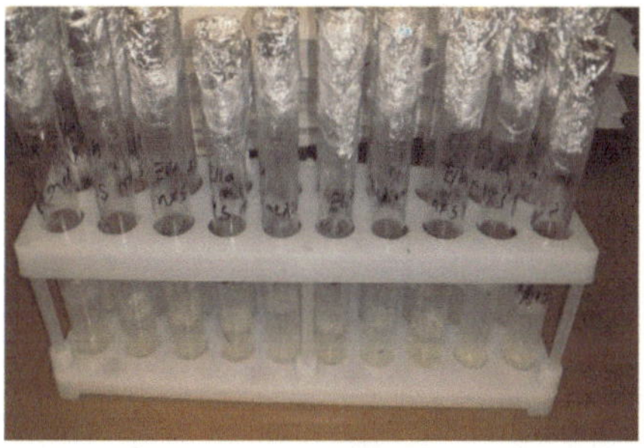

Figure 15.
*Stevia, chicory and bee balm (*Cambridge scarlet*),* Rhodiola krylovii *in solid media showing no growth sign after 4 weeks of culture.*

3.4 Preparation of cell suspension cultures

Two factors were used to determine the growth of callus in cell suspension (liquid media), and these are the use of the growth regulator kinetin and the absence of growth regulator – Kinetin in the media with the aid of microscopy results were obtained (see **Figures 16** and **17**).

3.5 Microscopy results of callus of chicory

a. The cells were viewed under the microscope after 1 week of culturing the callus

See **Figure 18**.

Figure 16.
Callus of chicory in liquid media of + kinetin and – Kinetin.

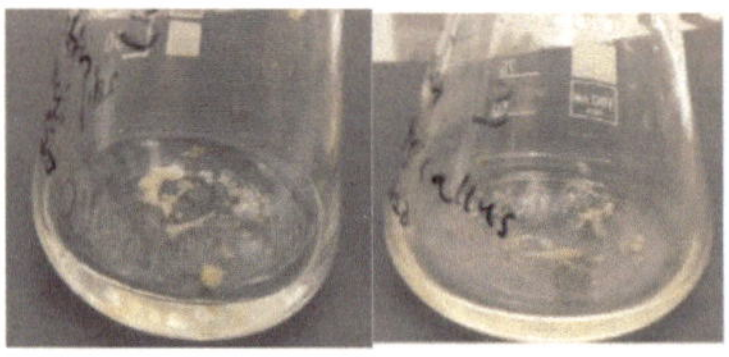

Figure 17.
Callus of chicory in the cell suspension.

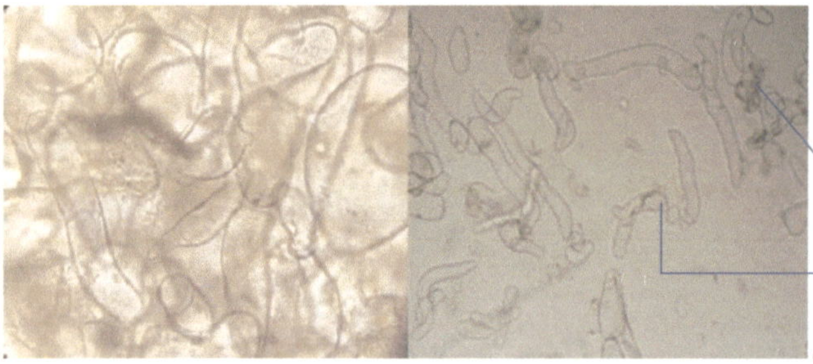

Figure 18.
(i) Chicory cells in liquid media (+kinetin). (ii) chicory cells in liquid media (– kinetin).

b. Cells after 2 weeks of Transferring into new Liquid culture media

See **Figure 19**.

The microscopy result after culturing the callus in an M and S liquid media of (−) and (+) Kinetin shows that the cells from the media with Kinetin developed more giant cells and several cells after 1 week of cultivation, while the cells in the media with Kinetin developed smaller and multiple cells. However, after 4-weeks, the cells from the media with Kinetin developed more giant cells in more significant numbers. In contrast, the cells from the media without Kinetin incurred contamination before the fourth week, and this result agrees with the findings of Usha et al. and Debnath [99, 103], which explains the necessity of administering a 14-day dark treatment for callus formation and the shoot regeneration of plants. The variation between the cells in media with Kinetin and media without Kinetin could be due to the addition of Kinetin, a growth regulator. It regulates the growth processes of cells by controlling cell division and cell differentiation affecting the plant's formation and regeneration process. Also, this correlates with the results of Usha et al. [99] that envaulted the

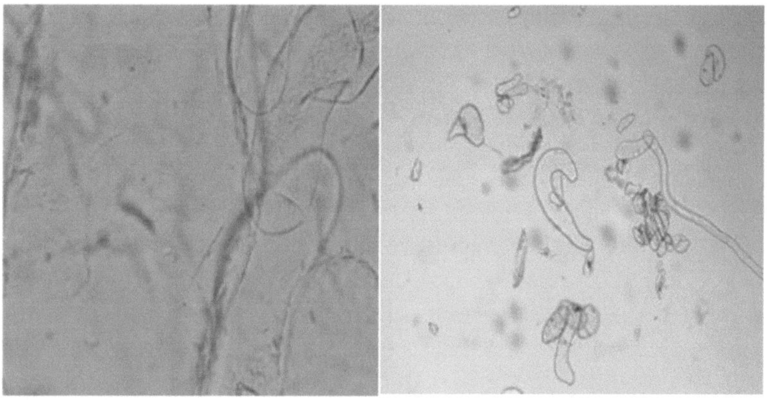

Figure 19.
(i) Chicory cells in liquid media (+kinetin) no contamination. (ii) chicory cells in liquid media (– kinetin) no contamination.c.) cells after 4 weeks of transferring into new cell suspension culture (liquid culture media) See ***Figures 20*** *and* ***21****.*

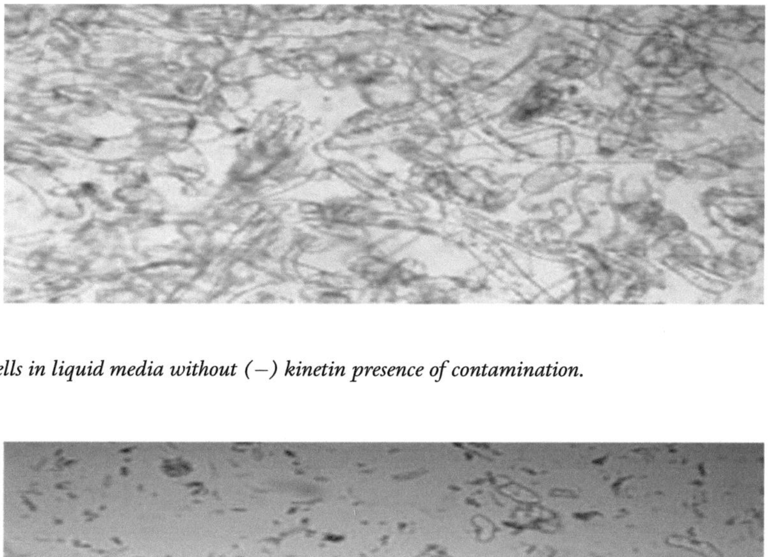

Figure 20.
(i) chicory cells in liquid media without (–) kinetin presence of contamination.

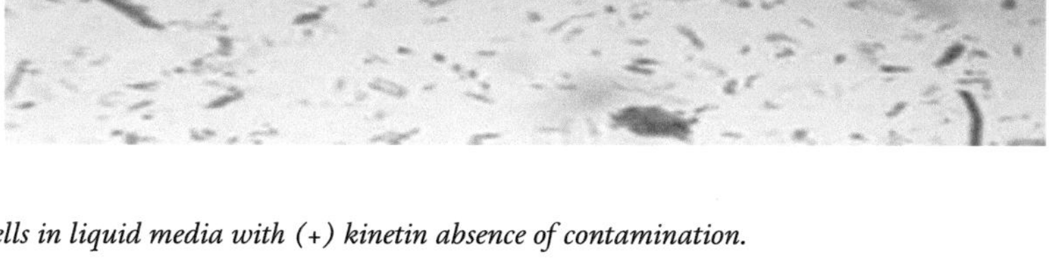

Figure 21.
(i) chicory cells in liquid media with (+) kinetin absence of contamination.

effect of plant regulators on plant cells. In addition, external factors, such as light intensity and temperature, could have affected the slow development of cells in the media with Kinetin. Since temperature and light play essential roles in plant growth, this agrees with the findings of Yildiz [57] elaborated on the influence of control equipment causing the culture room to be overheated resulting from temperature changes, which causes stress to plants enabling very low or no success in the growth rate.

3.6 Wet and dry matter determination

Two Factors – (I) Liquid culture media with (+) Kinetin and (II) Liquid culture media without (−) Kinetin were used to determine the outcome for wet and dry weight analysis of callus of chicory culture in liquid media.

3.6.1 Liquid culture media with kinetin (+ kinetin)

The result, as shown in **Figure 22**, can be deduced that the wet weight of samples A (0.81714), B (0.98178), and C (0.92866) had a close range in the values obtained for the wet weight and that of the dry weight for all samples showed relatively very low variation during drying hence, the dry weight between the samples is approximately the same. The sample with the lowest weight after drying was sample A (0.16974), next to it was sample C (0.17406), then B (0.17978). Thus, the upsurge in dry weight was due to cell splitting 14 and new material synthesis (see **Table 4**) [104, 105].

3.6.2 Liquid media without kinetin (− kinetin)

The result in **Figure 23** shows that the wet weight of samples A (0.97604), B (0.97378), and (1.16438) varied. Thus, making the dry weight fall within the same range. Nevertheless, A (0.17284) had the most negligible value after drying; next were B (0.17808) and C (0.18568), respectively. The increase in the wet weight before drying

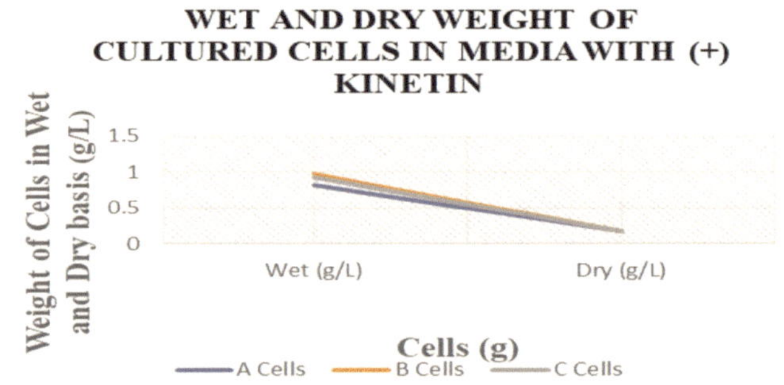

Figure 22.
Wet and dry matter determination of liquid culture media with kinetin (+ kinetin).

	Weight of Empty Filter paper	Volume of Sample	Weight of Paper + Wet Cells	Weight of Wet Cells	Weight of Dried Cells + Paper	Weight Dried Cells
	(g)	(L)	(g)	(g/L)	(g)	(g/L)
A	0.2098	0.005	0.8591	0.81714	0.2117	0.16974
B	0.2171	0.005	1.0252	0.98178	0.2232	0.17978
C	0.2137	0.005	0.9714	0.92866	0.2168	0.17406

Table 4.
Wet and dry matter determination of liquid culture media with kinetin (+ kinetin).

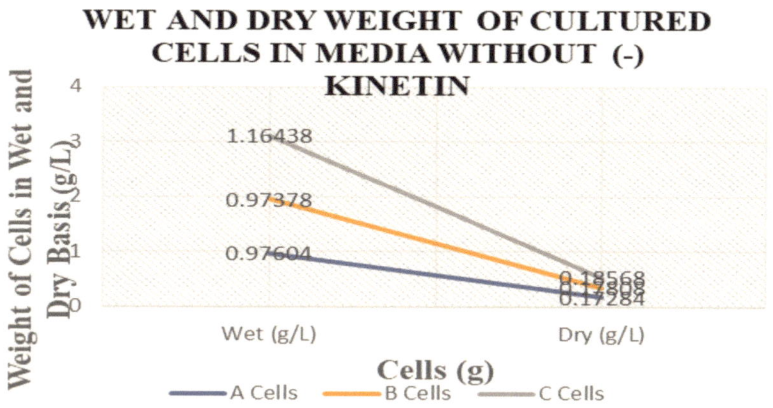

Figure 23.
Wet and dry matter determination of cells in liquid culture media without kinetin (− kinetin).

	Weight of Empty Filter paper	Volume of sample	Weight of Paper +Wet Cells	Weight of Wet Cells	Weight of Dried Cells + Paper	Weight Dried Cells
	(g)	(L)	(g)	(g/L)	(g)	(g/L)
A	0.2048	0.005	1.0170	0.9760	0.2138	0.1728
B	0.2096	0.005	1.0157	0.97378	0.2200	0.1781
C	0.2181	0.005	1.2080	1.1644	0.2293	0.1857

Table 5.
Wet and dry weight of cultured cells in media without kinetin (− kinetin).

resulted from external factors, leading to water absorption and turgor pressure of cells from the enlarged cell, which agrees with the study of Dale [106]. At the same time, the rapid decrease in moisture during drying could be as a result of the osmotic stress on the cells, which can cause moisture to escape from the cells [107] (see **Table 5**).

4. Technological part

4.1 Material balance

4.1.1 Raw materials for media preparation

Sucrose, solution of iron chelate, macronutrients, micronutrients, vitamins, Kinetin, 1-naphthalene acetic acid (NAA), dichlorophenoxyacetic acid, and distilled water.

1. **Sucrose** is a non-reducing disaccharide made of glucose and fructose linked via their anomeric carbons. It is a disaccharide formed by glucose and fructose units joined by an acetal oxygen bridge from the hemiacetal of glucose to the hemiketal of the fructose. It plays a role as an osmolyte, a sweetening agent, a human metabolite, an algal metabolite, a *Saccharomyces cerevisiae* metabolite, and an *Escherichia coli* metabolite. It is obtained commercially from sugarcane, sugar beet (*Beta vulgaris*), and other plants and is used extensively as a food and a

sweetener. The molecular formula of sucrose is **$C12H22O11$,** and the molecular weight is **342.3 g/mol.**

2. **Solution of Iron Chelate:** This compound helps maintain the pH in the culture media for plants and corrects an iron deficiency in plants. The molecular formula is **$C20H14FeN2O2^{+2}$,** and the molecular weight is **370.2 g/mol.**

3. **Macronutrients** are nutrients required by plants in more significant quantities, and they include: Nitrogen (N), Potassium (K), Calcium (Ca), Magnesium (Mg), Phosphorous (P), and Sulphur (S)

4. **Micronutrients** are those nutrients required by plants in fewer quantities, such as Chloride (Cl), Iron (Fe), Boron (B), Manganese (Mn), Zinc (Zn), Copper (Cu), Molybdenum (Mo), and Nickel (Ni)

5. **Vitamins**: Plants need vitamins for their growth, and examples of such vitamins include: Vitamin B, Vitamin C, and Vitamin E. Plants may derive specific benefits from applying these vitamins.

6. **Kinetin:** This is referred to as plant hormones known as cytokinin that helps to promote cell division and plant growth. Kinetin is an adherent part of the class of 6-aminopurines that is adenine carrying a (furan-2-ylmethyl) substituent at the exocyclic amino group. It is an adherent part of furans and a member of 6-aminopurines. It has a molecular formula of **C10H9N5O** and a molecular weight of 215.21 g/mol. It was shown to naturally exist in the DNA of organisms, including humans and various plants. In contrast, Kinetin is used in tissue cultures to produce new plants.

7. **1-Naphthalene acetic acid (NAA):** A synthetic plant hormone in the auxin family and is an ingredient in many commercial plants rooting horticultural products; it is a rooting agent used for the vegetative propagation of plants from stem and leaf cuttings. It is also used for plant tissue culture. It is an organic compound with the formula **C10H7CH2CO2H**. This colourless solid is soluble in organic solvents.

8. **Dichlorophenoxyacetic acid 2, 4-D** is a member of the phenoxy family of herbicides, which include: 2, 4-D is a synthetic auxin, which is a class of plant hormones, and by itself, it is often used in laboratories for plant research and as a supplement in plant cell culture media, such as MS medium. It was a primary ingredient in Agent Orange alongside its chemically similar relative, 2, 4, 5-T (2,4,5-trichloro phenoxy acetic acid). Its molecular formula is **C8H6Cl2O3** or **Cl2C6H3OCH2COOH,** and its molecular weight is 221.03 g/mol.

9. **70%Ethanol/Alcohol** is an ideal solution that is strong enough to reduce microbial contamination to an extent. It was used to sanitise the hands and working area before the media preparation [108, 109].

4.1.2 Process description

Stage 1: Preparation of Culture media.

The nutrient medium for growing chicory callus was 1 liter, which contained 30 g sucrose, 5 ml iron chelate solution, 50 ml macronutrient, 1 ml vitamins, 1 ml kinetin, 1 ml dichlorophenoxyacetic acid and 2.5 ml 1-naphthalene acetic acid (NAA).

Stage 2: Autoclaving.

The plant culture media was autoclaved at 121^0 C.

Stage 3: Callus Cultivation.

Cells of Chicory were inoculated in the sterile liquid culture media under controlled parameters of 16 hours light intensity phase at 24°C and 8 hours dark phase at 18°C for 1 week.

Stage 4: Separation.

A total of 5 ml of starter was pipetted into six centrifuge tubes, and centrifugation was done for 10 mins at 5000 rpm at 15°C. Thus, the aliquot was separated, and 5 ml of sterile water was added to the supernatant and centrifugation was repeated twice.

Stage 5: Drying.

Six (6) filter papers were dried in the oven for 9 hours at 60 °C, and the empty weight of the filter paper was obtained, and the weight of the filter with the sample was also obtained. The samples were later placed in the oven for 24 hours at 60°C. The weight of the dried sample on the filter was weighed. The fresh and dry weights were determined.

4.1.3 Process block diagram to produce dried chicory

See **Figure 24**.

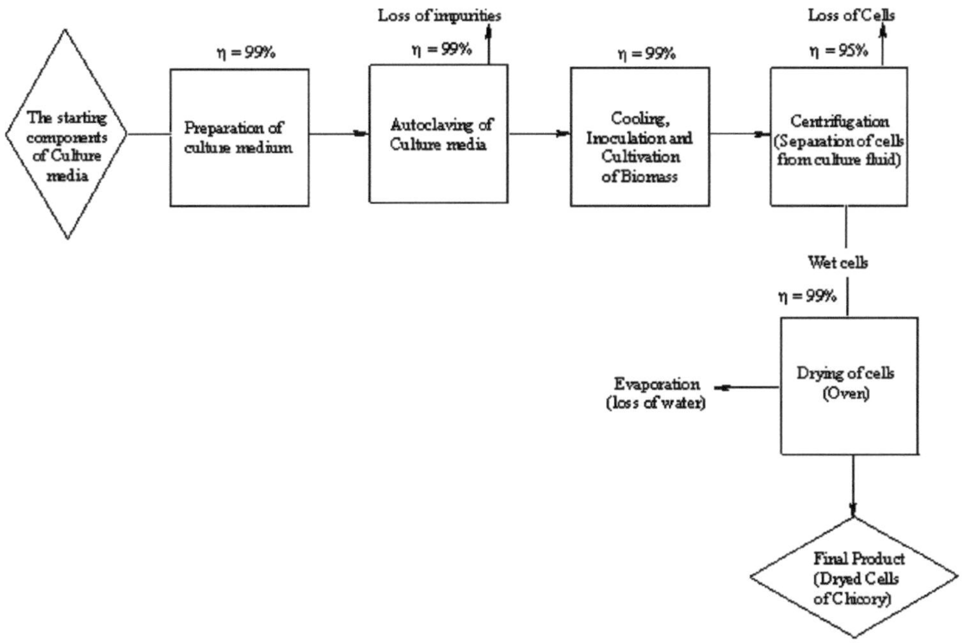

Figure 24.
Process block diagram to produce dried chicory cells.

4.1.3.1 Mass balance calculation

Characteristics of Final Product:

Form = Dried Cells Amount = 1 kg Purity = 98%.

Humidity = 1%.

Others (Impurities) = 1%.

Based on the production of 5 g of dry biomass per 1 L of culture medium (3% Sucrose).

The density of the nutrient medium - 1024 kg/m^3.

The yield of dry biomass, considering losses at the stages of cultivation, separation, and drying.

Overall yield (η) = 0.99 $*$ 0.95 $*$ 0.99 = 0.93%.

Quantity of Liquid culture media (3% Sucrose) considering losses.

Mass of culture medium = $\frac{1*0,001*1.024}{0,005*0.93}$ = 220.22 kg.

Considering losses (preparation and sterilisation) = 220.22/0.99/0.99 = 224.69 kg.

Mass of sugar needed = 224.69 $*$ 0.03 = 6.74 kg.

Iron chelate = 0.005 * 224.69 = 1.12 kg.

1-Napthalene acetic acid = 0.0025 * 224.69 = 5.12 kg.

Kinetin = 0.001 *224.69 = 0.56 kg.

Dichlorophenoxyacetic acid = 0.001 *204.80 = 0.22 kg.

Macronutrients = 0.05 *224.69 = 11.23 kg.

Micronutrients = 0.001*224.69 = 0.22 kg.

Vitamins = 0.001* 224.69 = 0.22 kg.

Total = 13.82 kg.

Quantity of water was calculated as:

Amount water needed = Amount of Liquid media – All solids (All media compositions) – Sugar Amount of Water needed = 224.69–13.82 – 6.74 (sugar) = 204.13 kg.

Summary of Raw Materials needed:

Sugar = 6.74 kg All solids = 13.82 Water = 204.13 kg.

Raw Materials:

4.1.3.2 First stage: Preparation of culture media

Process Yield (η) = 99%.

See **Table 6**.

Input				Output			
RAW MATERIALS	Content (%)	Weight (Kg)		PRODUCT	Content (%)	Weight (Kg)	
		Tech	100%			Tech	100%
Water	90.85		204.13	1) M&S Media		222.44	
Sugar	3		6.74	2) Losses		2.25	
Iron Chelate	0.5		1.12				
Napthalene acetic acid	0.25		0.56				
Kinetin	0.1		0.22				
Dichloro- phenoxyacetic acid	0.1		0.22				
			11.2				

Input				Output			
RAW MATERIALS	Content (%)	Weight (Kg)		PRODUCT	Content (%)	Weight (Kg)	
		Tech	100%			Tech	100%
Macronutrient	5	3					
Micronutrients	0.1	0.22					
Vitamins	0.1	0.22					
Total=		224.69		Total=		224.69	

Table 6.
Mass balance preparation of culture media.

4.1.3.3 Second stage: Autoclaving

Process Yield (η) = 99%.
See **Table 7**.

Input				Output			
RAW MATERIALS	Content (%)	Weight (Kg)		PRODUCT	Content (%)	Weight (Kg)	
		Tech	100%			Tech	100%
1) M&S Media		222.44		1) M&S Media		220.22	
				2) Losses		2.22	
Total=		222.44		Total=		222.44	
$\eta = 0.99$							

Table 7.
Mass balance for autoclaving.

4.1.3.4 Third stage: Cultivation of cells process: Yield (η) = 99%

Cultivation time – 336 h.
Air Consumption – 0.06 m^3/min Air Density – 1.39 kg/m^3.
Mass of Air = 336*60*0.06* 1.39 = 1681.34 kg.
See **Table 8**.

Input				Output			
RAW MATERIALS	Content (%)	Weight (Kg)		PRODUCT	Content (%)	Weight (Kg)	
		Tech	100%			Tech	100%
1) M&S Media		220.22		1) Culture fluid Including Biomass Water and Soluble/ Unsoluble components	0.2 0.8	219.42	5.28 214.14
2) Inoculum Including Media Plant cells	90 10	1.42	0.67 0.75	2) Air with Impurities and Losses		1683.56	
3) Sterile air		1681.34					
Total=		1902.98		Total=		1902.98	
$\eta = 0.99$							

Input				Output			
RAW MATERIALS	**Content (%)**	**Weight (Kg)**		**PRODUCT**	**Content (%)**	**Weight (Kg)**	
		Tech	**100%**			**Tech**	**100%**

Table 8.
Mass balance for cultivation of cells.

4.1.3.5 Fourth stage: Cell separation

Process Yield (η) = 95%.
See **Table 9**.

Input				Output			
RAW MATERIALS	**Content (%)**	**Weight (Kg)**		**PRODUCT**	**Content (%)**	**Weight (Kg)**	
		Tech	**100%**			**Tech**	**100%**
1) Culture fluid	0.2	219.42	5.28	1) Wet cells Including		5.02	1.00
Including Biomass	0.8		214.14	Dry substance Water		0.199	4.01
Water and Soluble/				Impurities		0.799	0.01
						0.002	
				2) Liquid phase		214.40	
Total=		**219.42**		**Total=**		**219.42**	
η = 0.95							

Table 9.
Mass balance for cell separation.

4.1.3.6 Fifth stage: Drying of cells

Process: Yield (η) = 99%.
See **Table 10**.

Input				Output			
RAW MATERIALS	**Content (%)**	**Weight (Kg)**		**PRODUCT**	**Content (%)**	**Weight (Kg)**	
		Tech	**100%**			**Tech**	**100%**
1) Wet cells Including Dry	19.92	5.02	1.00	1) Dried cells Dry	98	1.00	0.99
substance Water Impurities	79.88		4.01	substance Water	1		0.01
	0.20		0.01	Impurities	1		0.01
				2) Evaporated water			4.00
				3) Losses			0.01
Total=		**5.02**		**Total=**		**5.02**	
		η = 0.99					

Table 10.
Mass balance for drying cells.

5. Summary

The suitable sterilisation method for stevia seeds is method D, which gave 14.29% per cent contamination, while that of Chicory was method C, which gave 30% per cent contamination. However, for Bee balm and *Rhodiola krylovii,* the four methods used for sterilisation and cultivation yielded very high percentage contamination of 100%, which signifies that Bee balm and *Rhodiola krylovii* need a different method of sterilisation, and this can be concluded that each plant cell requires a specific sterilisation method for increasing yield and lowering microbial contamination. The use of growth regulators for callus formation and shoot regeneration should be used when obtaining the callus of medicinal plants. Hence, this can be seen in the cultivation process of callus of Chicory, Bee balm, and *Rhodiola krylovii,* which showed an increased growth rate and percentage contamination of 62.5% for both chicory and Bee balm and 75% while the callus of the same plants grown in media without Kinetin showed no growth. They had a very high percentage contamination of 100% for the same plant species (see **Figure 25**).

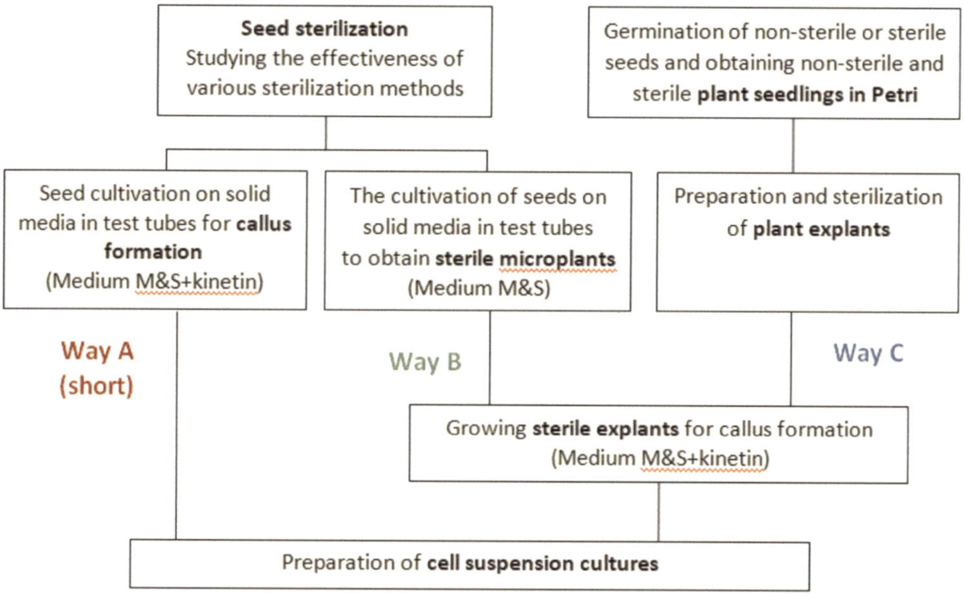

Figure 25.
Obtaining cell cultures of medicinal plants.

6. Conclusion

In a nutshell, the *in vitro* process of cultivating medicinal plants is a very tedious but delicate process that involves time, energy, proper care of the working area, and controlled environmental factors, such as temperature, humidity, moisture, and control equipment, which may alter the growing process and incur microbial contamination of plants. Thus, it was imperative to have studied the various sterilisation methods (A, B, C, and D) and the most effective method for the respective plants was used in this experiment since individual plant species have a particular method of sterilisation that can yield good results during *in vitro* propagation of these plants. Although the issue of contamination cannot be avoided entirely during the sterilisation and cultivation process of medicinal plants, it can be significantly reduced if the parameters (pH, concentration 3% and temperature of around 10°C and 11°C) of the sterilising agent, such as NaOCl, are maintained in the required form and the pH of the media should be adjusted to around 5.8 [110]. To improve these medicinal plants' production yield and the sterilisation, cultivation process of Bee balm and Rhodiola krylovii should be studied for further research. The bioactive compounds of Chicory, Stevia, Beebalm, and *Rhodiola krylovii* produced using methods C and D, which had less percentage contamination, should be compared to the bioactive compounds of another sterilisation and cultivation method that might be used in growing the same plants elsewhere.

Acknowledgements

I wish to acknowledge Professor Maxim A. Mironov for his intellectual input in this work and Osman N Kanwugu for his moral support and contribution in carrying out this research. Also, I would like to acknowledge the Department of Technology for organic Synthesis, Institute of Chemical Engineering, Ural Federal University Ekaterinburg, Russia where this research was successfully carried out.

Conflict of interest

The authors declare no conflict of interest.

Thanks

Special thanks to those who contributed to this research work and publication.

Author details

Torkwase Emmanuella Bulya*, Tatiana V. Glukhareva and Elena G. Kovaleva
Department of Technology for Organic Synthesis, Ural Federal University, Institute of Chemical Engineering, Ekaterinburg, Russia

*Address all correspondence to: nuellabulya@gmail.com

References

[1] Haberlandt G. Kulturversuche mit isolierten pflanzenzellen. Sber Akad Wiss Wein. 1902;**111**:69-92

[2] Thorpe TA. In: Bhojwani SS, editor. The Current Status of Plant Tissue Culture. In: Developments in Crop Science 19. Plant Tissue Culture: Applications and Limitations. Amsterdam: Elsevier; 1990. pp. 1-33

[3] Prinslooa G, Labuschagne J. Application of metabolomics in cultivation of medicinal plants. South African Journal of Botany. 2013;**86**:149-150

[4] Dorina P. Plant Growth and Cultivation. Springer Science: Methods in Molecular Biology; 2013

[5] Van-Wyk BE, Oudtshoorn BV, Gericke N. Medicinal Plants of South Africa. Pretoria: Briza Publications; 2013

[6] Cunningham AB. Management of medicinal plant resources. In: Seyani JH, Chikuni AC, editors. Proceedings of the 13th Plenary Meeting of AETFAT. Vol. 1. Zomba, Malawi, Limbe, Cameroon: Montfort; 1994. pp. 173-189

[7] Plants Mineral Nutrients. Methods and Protocols. Methods in Molecular Biology 953. Spring Protocols; 2013th edition, 2013. Humana: Humana Press; Oct 17, 2012. ISBN-13: 978-1627031516

[8] Cardoso JC, Oliveira M, Cardoso F. Advances and challenges on the in vitro production of secondary metabolites from medicinal plants. Horticultura Brasileira. 2019;**37**:124-132

[9] Víctor M. Loyola-Vargas and Neftalí Ochoa-Alejo, editors. Plant Cell Culture Protocols (Methods in Molecular Biology, 1815). 4th edn. Humana: Humana Press, 2018. ISBN-13: 978-1493985937

[10] Schippmann U, Leaman DJ, Cunningham AB. Impact of cultivation and collection on the conservation of medicinal plants: Global trends and issues. In: Jatisatienr A, Paratasilpin T, Elliott S, et al., editors. Conservation, Cultivation, and Sustainable Use of MAPs: A Proceeding of WOCMAP III: The IIIrd World Congress on Medicinal Aromatic Plants. Vol. 676. Chiang Mai, Thailand: ISHS, Leuven. Acta Horticulturae nr; 2005. pp. 31-44

[11] Lubbe A, Verpoorte R. Cultivation of medicinal and aromatic plants for specialty industrial materials. Industrial Crops and Products. 2011;**34**:785-801

[12] Ortega-Ramirez LA, Rodriguez-Garcia I, Leyva JM, Cruz-Valenzuela MR, Silva-Espinoza BA, Gonzalez-Aguilar GA, et al. Potential of medicinal plants as antimicrobial and antioxidant agents in food industry: A hypothesis. Journal of Food Science. 2014;**79**:2

[13] Borneo R, León AE, Aguirre A, Ribotta P, Cantero JJ. Antioxidant capacity of medicinal plants from the province of Córdoba (Argentina) and their in vitro testing in a model food system. Food Chemistry. 2009;**112**:664-670

[14] Oyenihi AB, Smith C. Are polyphenol antioxidants at the root of medicinal plant anti-cancer success. Journal of Ethnopharmacology. 2019; **229**:54-72

[15] Valdes ME, Ame MV, Bistoni MA, Wunderlin DA. Occurrence and bioaccumulation of pharmaceuticals in a fish species inhabiting the Suquia River basin (Cordoba, Argentina). Science Total Environment. 2014;**472**:389-396

[16] Baert JRA, Van-Bockstaele EJ. Cultivation and breeding of root chicory for inulin production. Industrial Crops and Products. 1992;**1**:229-234

[17] Velayutham P, Ranjithakumari BD, Baskaran P. An efficient in vitro plant regeneration system for Cichorium intybus L.–an important medicinal plant. Journal of Agriculture Technology. 2006; **2**:287-298

[18] Holloway L, Moynihan S, Abrams SA, Kent K, Hsu AR, Friedlander AL. Effects of oligofructose-enriched inulin onintestinal absorption of calcium and magnesium and boneturnover markers in postmenopausal women. The British Journal of Nutrition. 2007;**97**:365-372

[19] Shahvazian N, Shakiba M, Farahzadi MH, Mahmoodabadi F. Efficacy of chicory in decreasing serumferritin and liver enzymes in major beta thalassemia-patients. Iran Journal of Pediatrics Hematology Oncology. 2010;**1**:4-7

[20] Parnell JA, Raman M, Rioux K, Reimer, R.A. The potential role of prebiotic fibre for treatment and management of non-alcoholic fatty liver disease and associated obesity and insulin resistance. Liver International. 2012;**32**:701-711

[21] Creely S, McTernan P, Kusminski CM, Fisher FM, DaSilva NF, Khanolkar M, et al. Lipopolysaccharide activates an innate immune systemresponse in human adipose tissue in obesity and type 2diabetes. American Journal of Physiology. Endocrinology and Metabolism. 2007;**292**:740-747

[22] Landmann M, Kanuri G, Spruss A, Stahlb C, Bergheim I. Oral intake of chicoric acid reduces acute alcohol-induced hepatic steatosis in mice. Basic Nutritional Investigation. Nutrition. 2014;**30**:882-889

[23] Chatsudthipong V, Muanprasat C. Stevioside and related compounds: Therapeutic benefits beyond sweetness. Pharmacology & Therapeutics. 2009; **121**:41-54

[24] Singh S, Rao GS. The herbal sugar of 21st century. Sugar Technology. 2005;**7**: 17-24

[25] Scora RW. Interspecific relationships in the genus Monarda (Labiatae). University of California Publications in Botany. 1967;**41**:1-71

[26] Carnat AP, Lamaison JL, Remery A. Composition of leaf andpower essential oil from Monarda didyma L. cultivated in France. Flavour and Fragrance Journal. 1991;**6**:179-180

[27] Collins JE, Bishop CD, Deans SG, Svoboda KP. Composition of the essential oil from the leaves and flowers of Monarda citriodora var. citriodora grown in the United Kingdom. Journal of Essential Oil Research. 1994;**6**: 27-29

[28] Bishop CD, Thornton IB. Evaluation of the antifungal activity of the essential oils of Monarda citriodora var. citriodora and Melaleuca alternifolia on post-harvest pathogens. Journal of Essential Oil Research. 1997;**9**:77-82

[29] Bawa AS, Khanum F. Anti-inflammatory activity of Rhodiola roseae a second-generation adaptogen. Phytotherapy Research. 2009;**23**:102

[30] Skopinska-Rozewska E, Malinowski M, Wasiutynski A, Sommer E, Furmanowa M, Mazurkiewicz M, et al. The influence of Rhodiola quadrifida 50% hydro-alcoholic extract and salidroside on tumor-induced angiogenesis in mice. Polish Journal of Veterinary Sciences. 2008;**11**:97-104

[31] Kelly GS. Rhodiola rosea: a possible plant adaptogen. Alternatives Medicine. 2001;**6**:293-302

[32] Ganzera M, Muhammad I, Khan RA, Khan IA. Analysis of the marker compounds of Rhodiola rosea L. (Golden root) by reversed phase high performance liquid chromatography planta med. Chemical & Pharmaceutical Bulletin. 2001;**49**:465-467

[33] Petkov VD, Yonkov D, Mosharoff A, Kambourova T, Alova L, Petkov VV, et al. Effects of alcohol aqueous extract from Rhodiola rosea L. roots on learning and memory. Acta Physiologica et Pharmacologica Bulgarica. 1986;**12**:3-16

[34] Rohloff J. Volatiles from rhizomes of Rhodiola rosea L. Phytochemistry. 2002; **59**:655-661

[35] Li DJ, Lu B, Mo SR. Effects of rhodiola on hemodynamics in acute myocardial ischemia and heart failure in rats. Zhongguo Ying Yong Sheng Li Xue Za Zhi 2006; 22:291e2. Journal of Food and Drug Analysis. 2015;**23**: 359-369

[36] Maslova LV, Kondrat'ev B, Maslov LN, Lishmanov IB. The cardioprotective and antiadrenergic activity of an extract of Rhodiola rosea in stress. Eksperimental'naia i Klinicheskaia Farmakologiia. 1994;**57**:61-63

[37] Gao L, Wu C, Liao Y, Wang J. Antidepressants effects of Rhodiola capsule combined with sertraline for major depressive disorder: A randomized double-blind placebo-controlled clinical trial. Journal of Affective Disorders. 2020;**265**:99-103

[38] Vines G. Herbal harvests with a future: Towards sustainable sources for medicinal plants, Plant life International; www.plantlife.org.uk. 2004. ISBN: 1 904749-00-3

[39] Choi BH, Ohashi H. Generic criteria and an infrageneric system for Hedysarum and related genera (Papilinoideae–Leguminosae). Taxon. 2003;**52**:567-576

[40] Basiner T. Enumeratio monographica specierum gener-is Hedysari. Bulletin Cell Physics -Methods Implementation Science Academic St.-Petersb. 1845;**4**:305-315

[41] Fedtschenko BA. The genus Hedysarum. Acta Horti Petropolitani. 1902;**19**:185-342

[42] Canter PH. The catwalk of CAM – Fad and fashion in-complementary medicine. FACT. 2003;**8**:167-168

[43] Scalbert A, Manach C, Morand C, Rémésy C, Jiménez L. Dietary polyphenols, and the prevention of diseases. Critical Reviews in Food Science and Nutrition. 2005;**45**: 287-306

[44] Edwards R. No remedy in sight for herbal ransack. New Scientist (1971). 2004;**181**:10-11

[45] Cheng L, Han M, Yang LM, Li Y, Sun Z, Zhang T. Changes in the physiological characteristics and baicalin biosynthesis metabolism of Scutellaria baicalensis Georgi under drought stress. Industrial Crops and Products. 2018;**122**: 473-482

[46] Rahimi Y, Taleei A, Ranjbar M. Long-term water deficit modulates antioxidant capacity of peppermint (Mentha piperita L.). Sci. Hortic-Amsterdam. 2018;**237**:36-43

[47] Bistgani ZE, Siadat SA, Bakhshandeh A, Ghasemi-Pirbalouti A,

Hashemi M. Morpho-physiological and phytochemical traits of (thymus daenensis Celak.) in response to deficit irrigation and chitosan application. Acta Physiologiae Plantarum. 2017;**39**:231

[48] George EF. Plant propagation by tissue culture. Part 1- the technology. Edington: Exegetics Limited. 1993: 337-356

[49] Debergh PC. Effects of agar brand and concentration on the tissue culture medium. Physiologiae Plant. 1983;**59**: 270-276

[50] Kuria P, Demo P, Nyende AB, Kahangi EM. Cassava starch as an alternative cheap gelling agent for the in vitro micro-propagation of potato (Solanum tuberosum L.). African Journal of Biotechnology. 2008;**7**: 301-307

[51] Mbiyu M, Muthoni J, Kabira J, Muchira C, Pwaipwai P, Ngaruiya J, et al. Comparing liquid and solid media on the growth of plantlets from three Kenyan potato cultivars. American journal of. Experimental Agriculture. 2012;**2**:81-89

[52] Blumthaler M, Ambach W, Ellinger R. Increase in solar uv radiation with altitude. Journal of Photochemistry and Photobiology. B. 1997;**39**:130-134

[53] Körner C. Alpine plant life. Berlin: Springer Verlag; 2003

[54] Seigler D. Plant Secondary Metabolism. Dordrecht: Kluwer; 1998

[55] Kapoor S, Raghuvanshi R, Bhardwaj P, Sood H, Saxena S, Chaurasia OP. Influence of light quality on growth, secondary metabolites production and antioxidant activity in callus culture of Rhodiola imbricata. Journal of Photochemical and Photobiology B: Biology. 2018;**183**: 258-265

[56] Cardoso JC, Silva JAT. Micropropagation of Zeyheria Montana Mart. (Bignoniaceae), an endangered endemic medicinal species from the Brazilian Cerrado biome. In Vitro Cellular & Developmental Biology-Plant. 2013;**49**:710-716

[57] Yildiz M. The prerequisite of the success in plant tissue culture: High frequency shoot regeneration. Leva A, Rinaldi LMR, editors. Recent Advances in Plant in Vitro Culture. 2012. DOI: 10.5772/51097. ISBN 978-953-51-0787-3

[58] Preil W. General introduction: A personal reflection on the use of liquid media for in vitro culture a.K. In: Hvoslef-Eide WP, editor. Liquid Culture Systems for in Vitro Plant Propagation. Berlin, Germany: Springer Verlag; 2005. pp. 1-18

[59] Gaspar T et al. Plant hormones and plant growth regulators in plant tissue culture. In Vitro Cell Development Biology-Plant. 1996;**32**:272-228

[60] Debnath SC, McRae KB. An efficient adventitious shoot regeneration system on excised leaves of micropropagated lingon m berry (*Vaccinium vitis-idaea L.*). The Journal of Horticultural Science and Biotechnology. 2002;**77**:744-752

[61] Zhang YB, Jiang PP, Liu XY, Li WD. Mechanism and antibacterial activity of cinnamaldehyde against Escherichia coli and Staphylococcus aureus. Modern Food Science and Technology. 2015;**31**: 31-35

[62] Bustin S, Benes V, Garson J, Hellemans J, Huggett J, Kubista M. The MIQE guidelines: Minimum information for publication of quantitative real-time

PCR experiments. Clinical Chemistry. 2009;**55**:611-622

[63] Carazzone C, Mascherpa D, Gazzani G, Papetti A. Identification of phenolic constituents in red chicory salads (Cichorium intybus) by high-performance liquid chromatography with diode array detection and electrospray ionisation tandem mass spectrometry. Food Chemistry. 2013; **138**:1062-1071

[64] Pushparaj PN, Low HK, Manikandan J, Tan BKH, Tan CH. Anti-diabetic effects of Cichorium intybus in streptozotocininduced diabetic rats. Journal of Ethnopharmacology. 2007; **111**:430-434

[65] Tousch D, Lajoix AD, Hosy E, Azay-Milhau J, Ferrare K, Jahannault C. Chicoric acid, a new compound able to enhance insulin release and glucose uptake. Biochemical and Biophysical Research Communications. 2008;**377**:131-135

[66] Heimler D, Isolani L, Vignolini P, Romani A. Polyphenol content and antiradical activity of Cichorium intybus L. from biodynamic and conventional farming. Food Chemistry. 2009;**114**: 765-770

[67] Lin Z, Neamati N, Zhao H, Kiryu Y, Turpin JA, Abraham C, et al. Cichoric acid analogs as HIV-1 integrase inhibitors. Journal of Medicinal Chemistry. 1999;**42**:1401-1414

[68] Renard CMGC, Cre'peau MJ, and Thibault JF. Structure of the repeating units in the rhamno galacturonic backbone of apple, beet and citrus pectin. Carbohydrate Research. 1995;**275**:155-165

[69] Roberfroid MB. Inulin-type fructans: Functional food ingredients. The Journal of Nutrition. 2007;**137**:2493-2502

[70] Carbonell-Capella JM et al. High pressure processing of fruit juice mixture sweetened with Stevia rebaudiana Bertoni: Optimal retention of physical and nutritional quality. Innovative Food Science and Emerging Technologies. 2013;**18**:48-56

[71] Tavarini S, Angelini LG. Stevia rebaudiana Bertoni as a source of bioactive compounds: The effect of harvest time, experimental site and crop age on steviol glycoside content and antioxidant properties. Journal of the Science of Food and Agriculture. 2013; **93**:2121-2129

[72] Kinghorn ADS. The Genus Stevia. New York, USA: Taylor & Francis Inc; 2002. pp. 1-202

[73] Ricardo OG. Medicinal and aromatic plants—Industrial profiles stevia. In: Kinghorn DA, editor. Phytomedicine. Vol. 19. London: Taylor & Francis Inc.; 2005. p. 211

[74] Mazza G, Marshall HH. Geraniol, linalool, Thymol and carvacrol-rich essential oils from Monarda hybrids. Journal of Essential Oil Research. 1992;**4**: 395-400

[75] Rahimmalek M, Tabatabaei BES, Etemadi N, Goli SAH, Arzani A, Zeinali H. Essential oil variation among and within six Achillea species transferred from different ecological regions in Iran to the field conditions. Industrial Crops and Products. 2009;**29**:348-355

[76] Tohidi B, Rahimmalek M, Arzani A. Essential oil composition, total phenolic, flavonoid contents, and antioxidant activity of thymus species collected from different regions of Iran. Food Chemistry. 2017;**220**:153-161

[77] Fraternale D, Giamperi L, Bucchini A, Ricci D, Epifano F, Burini G,

et al. Chemical composition, antifungal and in vitro antioxidant properties of Monarda didyma L. essential oil. Journal of Essential Oil Research. 2006;**18**: 581-585

[78] Panossian A, Wikman G, Sarris J. Rosenroot (Rhodiola rosea): Traditional use, chemical composition, pharmacology and clinical efficacy. Phytomedicine. 2010;**17**:481-493

[79] Hsiu-Mei C, Hsin-Chun C, Chin-Sheng W, Po-Yuan W, Kuo-Ching W. Rhodiola plants: Chemistry and biological activity. Journal of Food and Drug Analysis. 2015;**3**:359-369

[80] Mao GX, Deng HB, Yuan LG, Li DD, Li YY, Wang Z. Protective role of salidroside against aging in a mouse model induced by D-galactose. Biomedical and Environmental Sciences. 2010;**23**:161-166

[81] Cuevas-Glory LF, Pino JA, Santiago LS, Sauri-Duch E. A review of volatile analytical methods for determining the botanical origin of honey. Food Chemistry. 2007;**103**: 1032-1043

[82] Igor J, Carlo IG, Tuberso MG, Dragan B. Composition of Sulla (Hedysarum coronarium L.) honey solvent extractives determined by GC/MS: Norisoprenoids and other volatile organic compounds. Molecules. 2010;**15**:6375-6385

[83] Shanidar SR. A Neanderthal flower burial in northern Iraq. Science. 1975; **190**:880-881

[84] Martens E, Demain AL. The antibiotic resistance crisis, with a focus on United States. Journal of Antibiotics. 2017;**70**:520-526

[85] Gupta E, Purwar S, Sundaram S, Rai GK. Nutritional and therapeutic values of Stevia rebaudiana: A review. Journal of Medicinal Plant Research: Planta Medica. 2013;**7**:3343-3353

[86] Mattarelli P, Epifano F, Minardi P, Di Vito M, Modesto M, Barbanti L, et al. Chemical composition and antimicrobial activity of essential oils from aerial parts of Monarda didyma and Monarda fifistulosa cultivated in Italy. Journal of Essential Oil Bear. 2017;**20**:76-86

[87] Ghabraie M, Vu KD, Tata L, Salmieri S, Lacroix M. Antimicrobial effect of essential oils in combinations against fifive bacteria and their effect on sensorial quality of ground meat. LWT - Food Science and Technology. 2016;**66**: 332-339

[88] Razak UNAA et al. In vitro micropropagation of Stevia rebaudiana Bertoni in Malaysia. Brazilian Archives of Biology and Technology. 2014;**57**: 23-28

[89] Skopinska-Rozewska E, Malinowski M, Wasiutynski A, Sommer E, Furmanowa M, Mazurkiewicz M, et al. The influence of Rhodiola quadrifida 50% hydro-alcoholic extractand salidroside on tumor-induced angiogenesis in mice. Polish Journal of Veterinary Sciences. 2008;**11**:97-104

[90] Chan SW. Panax ginseng, Rhodiola rosea and Schis and rachinensis. International Journal of Food Sciences and Nutrition. 2012;**63**:75-81

[91] Ajay GN. Cultivation of Medicinal and Aromatic Plants. Pune, India. Natural Products and Drug Discovery: Poona College of Pharmacy, Bharati Vidyapeeth Deemed University; 2018

[92] Datta SK, Chakraborty D, Janakiram T. Low-cost tissue culture: An

overview. The Journal of Plant Science Research. 2017;**33**:181-199

[93] Miranda GS, Souza SR, Amaro MOF, Rosa MB, Carvalho CA. Avaliação do conhecimento etnofarmacológico da população de Teixeiras. Revista de Ciências Farmacêuticas Básica e Aplicada: MG, Brasil; 2013

[94] Araújo KFM, Kerntopf MR, Oliveira DR, Menezes IRA, Brito Júnior FE. Plantas medicinais no tratamento de doenças respiratórias na infância: uma visão do saber popular. Revista da Rede de Enfermagem do Nordeste. 2012;**13**:3

[95] Bruning MCR, Mosegui GBG, Vianna CMM. A utilização da fitoterapia e de plantas medicinais em unidades básicas de saúde nos municípios de Cascavel e Foz do Iguaçu Paraná: A visão dos profissionais de saúde. Ciência & Saúde Coletiva. 2012;**17**:10

[96] Laher F, Aremu AO, Van SJ, Finnie JF. Concerns over Regulatory Practices in the Traditional Medicine Industry - Does Storage Affect the Biological Activity and Chemical Composition of Medicinal Plants. Research Centre for Plant Growth and Development, School of Life. Sciences. Private Bag X01, Scottsville 3209, South Africa: University of KwaZulu-Natal Pietermaritzburg; 2013

[97] Murashige T, Skoog F. A revised medium for rapid growth and bioassays with tobacco tissue cultures. Physiologia Plantarum. 1962;**15**(3):473-497

[98] Postek MT, Howard KS, Johnson AH, McMichael KL. Scanning Electron Microscopy: A Student's Handbook. Williston, VT: Ladd Research Ind., Inc; 1980

[99] Usha A, Asokan MV, Yaw LS, Dawn M, Samir CD. Liquid culture for efficient in vitro propagation of adventitious shoots in wild Vaccinium vitis-idaea ssp. minus (lingonberry) using temporary immersion and stationary bioreactors. Scientia Horticulturae. 2020;**264**:109-199

[100] Yildiz M, Er C. The effect of sodium hypochlorite solutions on in vitro seedling growth and shoot regeneration of flax (Linum usitatissimum). Die Naturwissenschaften. 2002;**89**:259-261

[101] Schull W. Temperature and rate of moisture intake in seeds. Botanical Gazette. 1920;**69**:361-390

[102] Racoppi F. Domestic Bleaches Containing Sodium Hypochlorite, Procter and Gamble, Italia SPA. Rome: Product Development Department; 1990

[103] Debnath SC. A two-step procedure for adventitious shoot regeneration from in vitro derived lingonberry leaves: Shoot induction with TDZ and shoot elongation using zeatin. Hort Science. 2005;**40**:189-192

[104] Sunderland N. Cell division and expansion in the growth of the leaf. Journal of Experimental Botany. 1960;**11**: 68-80

[105] Yildiz M, Ozcan S, Telci C, Day S, Özat H. The effect of drying and submersion pretreatment on adventitious shoot regeneration from hypocotyl explants of flax (Linum usitatissimum L.). Turkish Journal of Botany. 2010;**34**:323-328

[106] Dale JE. The control of leaf expansion. Annual Review of Plant Physiology. 1988;**39**:267-295

[107] Van Volkenburg E, Boyer JS. Inhibitory effects of water deficit on

maize leaf elongation. Plant Physiology. 1985;77:190-194

[108] https://pubchem.ncbi.nlm.nih.gov retrieved: May 25,2020

[109] US patent 6800482, Morikawa H, Takahashi M. Cultured cells of Australian laurel, Pittosporaceae and a method for culturing tissues by using said cultured cells. issued 2004-10-05. US 2002/0137206A1

[110] Jain SM, Gupta, editors. Protocols for Somatic Embryogenesis in Woody Plants. Copy right Springer; 2005. pp. 459-468

Chapter 6

Soilless Cultivation to Secure the Vegetable Demand of Urban and Peri-Urban Population

Duraisamy Kalaivanan, Govindan Selvakumar and Arockiasamy Carolin Rathinakumari

Abstract

Globally, more people live in urban areas than in rural areas, with 54% of the world's population residing in urban areas in 2014. It is estimated that by the year 2050, the world's population would increase to 10 billion, and close to 80% of human settlements would be concentrated in and around urban locations. This growing urban population will need access to healthy and nutritious food. To provide food to these urban-based inhabitants, vast areas of cultivable land might be required. But then again due to competition from sectors other than agriculture, researchers, experts, and planners are skeptical about the accessibility of such spread-out land parcels, particularly those around the prevailing and futuristic metropolitan agglomerations. This strong world-wide urbanization also puts a demand for producing vegetables in close proximity to the consumers. This gives us one of today's major challenges. Land, water, chemical fertilizers, and energy are vital resources for food production. Only 1% of freshwater found on the earth is accessible or available for human usage. Nearly 70% of that water is used in agriculture mainly for irrigation. Reserves of fertilizers, that are crucial for the production of food, are running out. With agricultural land becoming scarcer and the need for producing closer to or even in the cities to shorten the supply chain, not always the best soil can be chosen for producing crops. In this critical condition, we have to identify some alternatives to produce the vegetable crops without using soil medium in urban and peri-urban areas where rooftop/terrace space is available abundantly. When grown on the substrate, the quality of underlying soil is not a consideration, since plants do not root in the underlying soil; water and nutrients are delivered directly to the crop via the substrate. Substrate cultures can even take place without soil, for example, on concrete floors in buildings.

Keywords: soilless cultivation, cocopeat, vegetables, nutrient management, and urban space

1. Introduction

Soil is usually the most available growing medium for all kinds of plants. Almost all of the vegetables we find on grocery store shelves are produced either directly or indirectly in open field soils. In general, soil serves two basic purposes—it acts as a reservoir to retain nutrients and water, and it provides physical support for the plant through its root system [1]. A well-drained, pathogen-free field soil of uniform texture is the least-expensive medium for plant growth, but the soil does not always occur in this perfect package [2]. Existing levels of abiotic and biotic stresses in soil severely affects agricultural and horticultural production. Some soils are poorly textured or shallow and provide an unsatisfactory root environment because of limited aeration and slow drainage. Pathogenic organisms are a common problem in field soils. On the other side, the shrinking of agricultural land due to continuous urbanization and industrialization also affects the total agriculture and horticulture production [1]. Strong worldwide urbanization also puts a demand for producing vegetables in close proximity to the consumers. When adverse conditions are found in soil and reclamation is impractical, some form of an alternate method of cultivation without soil may be justified. Soilless cultivation is another way of growing agricultural and horticultural crops. The recent scientific invention proved that it is also possible to produce crop plants without soil, *i.e., soilless culture* [3–5].

Presently, many countries are focusing special attention towards soilless cultivation, *i.e.,* hydroponics, aeroponics, and other substrates medium, such as cocopeat and compost [6]. Subtropical countries, such as India, aeroponics or cocopeat substrate-based cultivation is ideal as water is precise input for us. Besides cocopeat, several other substrates *viz.,* sand, rockwool, vermiculite, expanded clay granules, perlite, zeolite, and pumice could be used alone or by mixing with other organic or inorganic substrates as a medium for growing crops. Among all available organic substrates, peat moss mined from the earth is the highly used growing medium in horticulture, particularly in the nursery sector. However, peat moss is a limited resource with great demand, and the extraction of peat causes negative impacts on the environment. Most of the soilless substrates have superior hydraulic and physical features than those of soil and also permit synchronized optimization of oxygen and water availabilities for plant growth. Nutrient availability to plants can be better managed in the soilless system of cultivation than in most soils. With the help of the soilless system of cultivation, it is possible to minimize or reduce the discharge of dissolved ions, pesticide residues, etc., to the freshwater bodies which in turn prevent environmental pollution. Further, the carbon and water footprint can be reduced by practicing the soilless system of cultivation in limited available land on the earth. Soilless substrate-based cultivation improves water and nutrient use efficiencies when compared to the soil which ultimately minimizes carbon and water footprint.

Soilless culture is rapidly gaining momentum and popularity and is one of the fastest-growing sectors of agriculture. There has already been a great deal of buzz throughout the scientific community for the potential to use soilless culture in future food production. Soilless culture could well dominate food production in the future. The application of these systems is likely to increase close to existing cities as well as in mega-cities worldwide in the near future. To meet the growing demand for soilless culture technology, ICAR-Indian Institute of Horticultural Research, Bengaluru has standardized a simple and low-cost production technology, including nutrient formulations for open and polyhouse soilless cultivation of most commonly consumed vegetables *viz.,* tomato, chilli, cabbage, cucumber, French bean, garden peas, ridge gourd and leafy vegetables and few exotic vegetables, such as zucchini and colour cabbage using Arka Fermented Cocopeat (AFC) as substrate. Therefore, the production of vegetables under soilless culture using Arka Fermented Cocopeat and IIHR standardized nutrient solution namely Arka Sasya Poshak Ras may be practiced for meeting the demand of the urban population. Using this technology, urban and peri-urban people can grow their choice of vegetables to meet their daily vegetable requirements [7]. This chapter provides the reader with an understanding of the availability of various soilless media, specifications of the ideal substrate, suitable vegetable crops for soilless culture, how to grow vegetables under soilless culture, and nutrient and water management. This chapter is ideal for agronomists, horticulturalists, greenhouse and nursery managers, extension specialists, and people involved with the production of plants particularly vegetable crops under soilless culture.

2. Soilless culture

Soilless culture is a method of growing plants without soil. In this method of cultivation, plants are grown by providing nutrients, water, and physical support in a container. Soilless culture is normally called water or solution culture, the technique was firstly termed by W.F. Gericke as hydroponics (water working) in the 1930s [8]. Several workers use the term hydroponics to mention the systems that include some kind of organic or inorganic substrates to support the plant physically and to hold water in its inert matrix. The hydroponics method of cultivation has been used every now and then in the world as a profitable business of growing vegetable, flower, ornamental and medicinal plants. Because of the availability of various types of substrates along with scientific advancements, soilless culture has entered into the viable commercial stage. It supplies fresh vegetables in countries with limited arable land as well as in small countries with dense populations. Plants grown in hydroponics or soilless culture had consistently superior quality, high yield, rapid harvest, and high nutrient content.

3. Container growing of vegetable crops

Soilless culture in bags, pots, or troughs with a lightweight medium, *i.e.,* cocopeat is the simplest, most economical, and easiest to manage of all soilless systems. There are different types of containers are available, *i.e.,* long wooden troughs in which one or two rows of plants are grown, polyethylene bags, or rigid plastic pots containing one to three plants. In the bag or pot system, the solution is not recirculated. The most common types of media used in containerized systems of soilless culture are peat-lite, or a mixture of bark and wood chips. Bag or pot systems using bark chips or peat-lite are in common use [9]. Drain holes should be provided in the base of the containers to drain out any excess water or flush out any excess nutrient solution from the container.

4. Vegetable crops suitable for cultivation under soilless culture

The existence of a diverse climate in India ensures the availability of all types of fresh vegetables. India stands second in vegetable production in the world, after China. As per National Horticulture Database (Second Advance Estimates) published by National Horticulture Board, during 2019–2020, India produced 191.77 million metric tonnes of vegetables. The area under vegetable cultivation is 10.35 million hectares. The global area under soilless cultivation of vegetables is 95,000 ha only. This is a very meagre area at the world level when compared to an area under

Type of vegetable crops	Name of the vegetable crops
Transplanted vegetables	Tomato, brinjal, chilli, onion, cabbage, cauliflower, and broccoli
Direct sown vegetables	Okra, zucchini, cucumber, ridge gourd, bottle gourd, spine gourd, radish, beetroot
Perennial vegetables	Drumstick, curry leaf, chekkurmanis and agathi
Leafy vegetables	Amaranthus, palak, and lettuce
Spice crops	Coriander and fenugreek
Legume vegetables	French bean, garden peas, Dolichos, cowpea, and yard long bean
Source: Kalaivanan et al. [10].	

Table 1.
List of vegetable crops that can be grown successfully under soilless culture.

soil-based cultivation of vegetables. There is a range of limitless options in soilless culture regarding the type of vegetable crops to be grown. The list of suitable vegetable crops under different groups for growing in both open-field and polyhouse soilless culture conditions is given in **Table 1**.

5. Home garden structures

Based on the space available in terrace or rooftop of home two types of gardens can be adopted *viz.,* 1. terrace garden and 2. vertical garden. If a terrace or back yard or front yard space is available the vegetables and medicinal herbs can be grown in the accessible area. In the recent past, urban areas have become thickly populated and society prefers to live in apartment-type living places owing to its many advantages. For apartment dwellers, space is the limiting factor to grow their vegetables and medicinal herbs. However, every apartment has a utility area, where vegetables and medicinal herbs can be grown by adopting the vertical garden structures. In both types of gardens, vegetable and medicinal herbs are grown in pots or grow bags or rectangular trays.

6. Terrace garden structure

Terrace garden can be two models i. open garden and ii. shade net garden. In an open garden, containers are placed on the terrace, and vegetables and medicinal herbs are grown. Hence, the investment is only on containers, growing media, seeds, crop production and protection chemicals, and home garden tools. In the case of a shade net garden, a shade net is installed and crops are grown inside the shade net. The investment is Rs 100/square feet in addition to the above-mentioned investment. However, the shade net garden protects the plants from pests and diseases to a greater extent, reduces the use of crop protection measures, and the crops and produce are much healthier as they are grown under protected conditions.

7. Shade net garden

Installation of shade net is very simple and can be done by any local artisans (**Figure 1**). It requires galvanized pipes (G.I.) of 60 mm diameter ("B" Class),

Figure 1.
Shade net garden.

fasteners, and an agro shade net (50%). The length and width of the shade net can be any size based on the area available and the height will be 8½ feet. The G.I pipes are grouted to the terrace if it is to be on a terrace or can be fixed on the ground with proper concrete foundation if it is to be on an open yard. The space between two adjacent G.I. columns is 10 feet. A simple door is required to be provided at any one convenient place of the structure. The dimension of the door is 1.2 × 1.8 m (W × H). The entire G.I structure is covered with a 50% agro shade net with fasteners. The containers, such as grow bags, pots, and rectangular trays, can be placed inside the shade net.

8. Vertical garden structure

The vertical home structure is designed considering (i) size suitable for terrace/utility area, (ii) to grow vegetables consumed by a family on daily basis, (iii) pots suitable for respective vegetables/leafy vegetables/flowers/medicinal plants, (iv) structure suitable for handling in terms of the height of reach, mobility, the requirement of light available to all the pots, and (v) effective utilization of maximum area for growing plants. The vertical garden structure has three major substructures *viz.,* (i) base frame, (ii) main central support and (iii) supports for pots/grow bags (**Figure 2**). Main centre support is a rectangular shape frame/tube anchored to the base frame with necessary supports. Support for pots/grow bags are fabricated suitable for different pot sizes and shapes and fitted at four different height levels. Heavy-duty nylon caster wheels are fitted at the bottom of the base frame for the mobility of the vertical garden structure. The selection of pot size is based on the growing media requirement to facilitate proper growth during the crop period.

The vertical garden has four height levels and the topmost level was decided based on the maximum reach of a normal human being hand reach. Vegetable crops that grow a height of higher than 2 feet (tomato, chilli, brinjal, peas, etc.,) are placed in

(a) (b)

Figure 2.
Vertical garden.

the bottom-most level of the vertical garden structure. Leafy vegetables (palak, amaranthus, coriander, etc.) that grow to a height of about one foot are placed above the bottom layer. Medicinal crops or again leafy vegetables are placed above the second bottom layer. Flowers are placed at the topmost level of the structure which would give aesthetic look.

9. Characteristics of growth medium (substrates) used in soilless cultures

9.1 Technical specifications for substrates

According to [11–13] substrates must have the following properties:

- Inert (no reaction with the nutrients)
- pH neutral
- Porous
- Low density
- Hydrophilic
- There should not be any radioactive pollutants and heavy metals in substrates

- As much as possible the substrate should be usable in natural form without any additional processing
- The substrate can either be obtained by mining from nature or otherwise produced in the industry
- It should have constant quality without much change particularly in physical properties during use
- Substrate should have a lifetime of a minimum of 3 years
- The substrate should be easy to handle and use
- The cost of the substrate should be low
- The nature of the substrate should be either biodegradable or destroyed without causing any environmental risk
- It should not undergo any structural change during repeated sterilization
- The substrate must be free from pest and disease-causing agents/pathogens.

Substrates, such as rockwool, cocopeat, clay granulates, pumice, sand, Irish peat, and perlite, are able to meet the above specifications [14, 15].

9.2 Substrate categories

Ideal substrate should fulfil four important roles *viz.,* (i) should act as a reservoir for plant nutrients, (ii) should hold enough water, (iii) provide good aeration, and (iv) physical support for good plant growth and development. Only a few substrates that are available in the market generally support all the four functions as mentioned above but at a very limited level. However, some of the soilless media may not support all the functions, i.e., sand gives good physical support and

Authors reference	Country	Area in ha	Media/ system	Key crops grown
Hassall *et al.* [16]	Spain	4000	Rockwool, sand, perlite	Cucumber, capsicum, tomato, lettuce
Hassall *et al.* [16]	Netherlands	10,000	Rockwool	Strawberry, tomato, cucumber, lettuce, cauliflower, muskmelons, gerbera, chrysanthemum, carnation
Jiang *et al.* [17]	China	1250	Rockwool, NFT, DFT	Carnation, roses, chrysanthemum, tomato, cucumber, lettuce
Donnan [18]	France	1000	Rockwool	Capsicum, tomato, cucumber, cut flowers
Bradley *et al.* [19]	Canada	2000	Rockwool and perlite	Cucumber, capsicum, tomato,

Table 2.
Various soilless culture media and crops grown.

aeration but is very poor in nutrient and water supplying capacity. Various growing media and crops grown in different countries are given in **Table 2**. Based on the source or origin of substrates, soilless growing media may be classified into two groups namely organic and inorganic medium. Peat moss, wood residues, sawdust, barks, and cocopeat are some of the widely used organic substrates in soilless cultivation. Other substrates, such as perlite, sand, vermiculite, calcined clays, pumice, and rockwool are most commonly used inorganic substrates in hydroponics or soilless cultivation.

10. Arka Fermented Cocopeat substrate

An ideal potting medium for vegetable crops must be well aerated and porous, hold sufficient moisture, have adequate drainage, and must provide adequate nutrients to the plants. Among all substrates, cocopeat is the one that retains moisture, stores, and releases nutrients to roots over an extended period of time for enhancing plant growth. Therefore, it is considered an ideal soilless growing media for vegetable crops. In this connection, the technology for conversion of raw coir pith into fermented cocopeat has been standardized at ICAR-Indian Institute of Horticultural Research, Bengaluru and released as a product called Arka Fermented Cocopeat (AFC). Arka Fermented Cocopeat is developed by the solid-state fermentation of raw coir pith, by employing a fungal consortium and enriched with the Arka Microbial Consortium comprising of N fixing, P and Zn solubilizing, and plant growth-promoting microbes could be a potential substrate for soilless cultivation of vegetables, flowers, and medicinal crops, etc. Arka Fermented Cocopeat is very popular and used as a growing media in the nursery for raising seedlings of various vegetable crops and rootstocks of different fruit crops. However, it has not been evaluated as a growing media for the cultivation of vegetables under soilless conditions. Therefore, a series of experiments on soilless cultivation of different vegetables were conducted at ICAR-IIHR to study the suitability of Arka Fermented Cocopeat (AFC) as substrate along with commercial cocopeat and soil. The results revealed that the substrate AFC recorded better yield and quality in vegetable crops compared to commercial cocopeat and soil. Arka Fermented cocopeat (AFC) alone or AFC + vermicompost or AFC + vermicompost/FYM/compost are also the best substrate combination for growing vegetable crops under soilless cultivation.

Raw coir pith fermented cocopeat

11. Nutrient management in soilless vegetable cultivation

Seventeen nutrient elements are considered essential for the growth and development of any living plant on the earth. The absence of anyone essential nutrient will make it difficult for the growth of plants and will not allow the plant to complete its life cycle. Further, the role of essential nutrients cannot be played or replaced by any other nutrients. In soilless culture or hydroponics, the nutrients which are considered essential should be supplied in the form of nutrient solution. Mostly C, H, and O are taken by the plants from water and CO_2 in the air. Remaining essential nutrients *viz.*, P, K, Ca, Mg, S, Fe, B, Cu, Zn, Mn, Mo, and Cl need to be supplied to the plants by the growers in the form of the nutrient solution prepared from soluble inorganic salts. Most of the growing media used to have a smaller quantity of these mineral nutrients but they should not be taken into account while formulating fertilizer schedule because they are either very little when compared to the requirement of plants or they may not be readily available form to plants. The main route to attain success in the soilless culture of vegetables depends on the ideal management of plant nutrients through the supply of nutrient solutions prepared from soluble inorganic salts. It is also possible to prepare a nutrient solution by buying a ready mix of all essential nutrients. There are several formulations of nutrient solutions available in the literatures. Nevertheless, most of them are empirically based. **Table 3** comprises some of them.

Proper nutrition factors, such as pH level, electrical conductivity (EC), the types of nutrition, the composition of nutrients irrigated, and so on are the key factors to improve the quality and yield of vegetables. Vegetable crops can be grown organically by mixing organic manures, such as FYM, compost (kitchen waste compost, city compost), and vermicompost, with substrate cocopeat @ 1:1:1 ratio which will take care of the nutrient requirement of the plants. Vermicompost @100 g/plant should be applied at monthly intervals. Decomposed kitchen waste can also be applied. ICAR-IIHR standardized

Nutrient	Cooper [20]	Steiner [21]	Hewitt [22]	Hoagland & Arnon [23]
			mg L^{-1}	
N	200–236	168	168	210
P	60	31	41	31
K	300	273	156	234
Ca	170–185	180	160	160
Mg	50	48	36	34
S	68	336	48	64
Fe	12	2–4	2.8	2.5
Cu	0.1	0.02	0.064	0.02
Zn	0.1	0.11	0.065	0.05
Mn	2.0	0.62	0.54	0.5
B	0.3	0.44	0.54	0.5
Mo	0.04	Not considered	0.04	0.01

Table 3.
Concentration ranges of essential mineral elements according to various authors.

nutrient solution (Arka Sasya Poshak Ras) may be practiced for meeting the nutrient requirement of the plants under cocopeat-based soilless cultivation. Arka Sasya Poshak Ras is a liquid nutrient formulation (comprising solutions A and B) is a unique blend of the macro and micronutrients which are well balanced to support the growth of vegetables.

12. How to use Arka Sasya Poshak Ras

- It is suitable for most commonly used vegetables (tomato, chilli, cabbage, zucchini, cucumber, ridge gourd, French bean, peas, cowpea, Dolichos, etc.) and leafy vegetables (amaranthus, coriander, palak, etc.)
- One litre each of nutrient solution *viz.*, A and B can be diluted with 200 litres of water (5 ml/litre) and applied @ 200 ml per plant (tomato, chilli, brinjal, cabbage, cucumber, ridge gourd, and zucchini).
- For leafy vegetables, 3.5 ml of each nutrient solution A and B may be diluted in 1 litre of water and applied @ 600 ml per bag of size 4 × 1 × 1 feet.
- For peas, beans, Dolichos, and cowpea, 4.0 ml of each nutrient solution A and B may be diluted and applied @ 600 ml per bag of size 4 × 1 × 1 feet.
- The frequency of nutrient solution application is two times per week starting from the 10th day of transplantation up to 30 days from the date of sowing or transplanting and three times per week thereafter.

13. Optimum pH of the nutrient solution

pH regulates/controls the availability of most of the essential plant elements in a nutrient solution. The nutrient solution pH between 5.8 and 6.5 is considered as most optimal. Higher or lower nutrient solution pH than the suggested range for individual crops, the nutrient deficiencies will become apparent or toxicity symptoms will grow.

14. Electrical conductivity of the nutrient solution

Similar to pH, electrical conductivity (EC) is one of the most important properties of nutrient solutions. The EC level between 1.5 and 2.5 dS/m is considered ideal for hydroponics/soilless culture. The strength of the nutrient solution strictly depends on the EC level of the solution. The total concentration of the solution is only indicated by the EC and not the specific nutrient components. Too high or too low EC level in nutrient solution may create salinity problems or the supply of some nutrients to the crop may be insufficient. Higher EC will not allow nutrient absorption to take place due to osmotic pressure and lower EC severely affects plant health and yield. However, among different species, the yield response of the plants may vary widely with respect to the EC level of the nutrient solution. So, the terms "too low" and "too high" need to be quantitatively defined for each cultivated plant species based on experimental results. When plants take up nutrients and water from the solution, the total salt concentration, i.e., the EC of the solution changes. Freshwater must be added If the EC is higher than the recommended range. Add nutrients if the EC is lower in the nutrient solution.

DOI: http://dx.doi.org/10.5772/intechopen.102695

15. Production technology for soilless cultivation of vegetables

The production technology for soilless cultivation of zucchini, colour cabbage, chilli, coriander, cucumber, French bean, peas, and tomato on Arka Fermented Cocopeat under open as well as in protected conditions has been standardized at ICAR-Indian Institute of Horticultural Research, Bengaluru. The results of most of the experiments conducted with different vegetable crops in grow bags under open-field and polyhouse soilless culture indicated that the plants grown in soilless culture recorded higher yield and better quality, particularly in mineral nutrient content compared to soil-grown plants. This technology would be highly suited for urban and peri-urban vegetable cultivation for meeting the food security in cities. This particular technology has already been popularized through various training programmes, exhibitions, magazines, and media. Many growers have already started adopting IIHR soilless culture technology in the cultivation of vegetables using AFC as a substrate.

16. Open-field vs polyhouse soilless cultivation of vegetables

Between open and polyhouse soilless cultivation, the highest yield and better fruit quality were recorded with zucchini, chilli, coriander, cucumber, French bean, peas, and tomato with open conditions. However, colour cabbage recorded maximum head weight and highest yield in polyhouse soilless cultivation. Similarly, brinjal also recorded higher yield in polyhouse than in open-field soilless culture because of better control of pests, particularly brinjal shoot and fruit borer. Pest and disease management was easier in polyhouse than in open-field soilless culture [24].

16.1 Zucchini

Between open and polyhouse soilless cultivation of zucchini, the highest stem diameter (35.2 mm), maximum fruit length (23.2 cm), fruit girth (42.9 mm), fruit weight (315.4 g), and yield (5.27 kg/plant and 65.8 t/ha) were recorded with open conditions [25, 26]. However, the maximum plant height (80.6 cm), number of leaves (47.4), number of fruits (22.3), and total plant dry biomass (144.7 g/plant) were recorded with polyhouse conditions.

16.2 Red cabbage

When open and polyhouse soilless cultivation of red cabbage were compared, the maximum plant height (25.71 cm), head diameter (33.7 cm), head length (12.9 cm), average head weight (817.8 g/plant), and yield (45.43 t/ha) was recorded with polyhouse conditions [27].

16.3 Chilli

Open field soilless cultivation outperformed polyhouse cultivation in almost all the parameters (number of fruits (228), fruit length (11.8 cm), fruit girth (10.3 cm), average fruit weight (5.68 g), and yield (1.29 kg/plant)) recorded during the course of the experiment except plant height [28].

16.4 Coriander

The performance of coriander under open-field soilless culture was found to be better than polyhouse soilless culture [28].

16.5 French bean

In French bean, open field soilless cultivation outclassed polyhouse in stem diameter (11.2 mm), number of branches (6.14), number of pods (42.74), pod length (15.04 cm), pod girth (7.23 mm), and pod yield (286.4 g/plant)) recorded during the course of the experiment except for plant height [29].

16.6 Garden peas

Best nutrient scheduling found in open-field conditions recorded better growth and yield in garden peas under polyhouse also. Between soil and cocopeat, soil recorded maximum growth and better yield compared to cocopeat [30].

17. Substrate effects on growth, yield, and quality of vegetables

With respect to different substrates studied, zucchini, chilli, coriander, cucumber, and tomato raised on Arka Fermented Cocopeat registered better growth and yield than soil. However, colour cabbage and peas recorded better growth and yield with soil. French bean plants recorded on par yield with both soil and soilless substrate.

17.1 Zucchini

Zucchini plants recorded maximum plant height (54.7 cm), stem diameter (35.2 mm), number of leaves (39.3), total plant dry biomass (139.8 g/plant), number of fruits (16.8), fruit length (23.2 cm), fruit girth (42.9 mm), fruit weight (315.4 g) and yield (5.27 kg/plant and 65.8 t/ha) when the plants raised on Arka Fermented Cocopeat compared to soil (3.70 kg/plant and 46.3 t/ha) [25, 26].

17.2 Colour/red cabbage

Among the substrates, soil registered maximum stem diameter (24.9 mm), number of leaves (28.3), head diameter (36.8 cm), head length (13.7 cm), average head weight (977.8 g), and yield (54.32 t/ha) in red cabbage compared to Arka Fermented Cocopeat (817.8 g and 45.43 t/ha, respectively). Nevertheless, AFC recorded maximum plant height (25.7 cm) than soil (24.5 cm) [27].

17.3 Chilli

In grow bags, chilli raised on Arka Fermented Cocopeat registered maximum number of fruits (232), fruit length (11.8 cm), fruit girth (10.3 mm), average fruit weight (5.68 g), and yield (1.29 kg/plant) compared to soil (1.02 kg/plant) [28].

17.4 French bean

Plants grown in AFC and soil (41 pods, 6.83 g pod weight, 283 g/plant, and 19.97 t/ha) were recorded on par yield with each other. Most of the macro and micronutrient concentrations in French bean pods were found to be higher in soilless plants than in those grown in soil [29].

17.5 Garden peas

Between soil and cocopeat, soil recorded maximum growth and better yield compared to cocopeat. The results showed or indicated that the soil is found to be more suitable for peas followed by cocopeat. However, most of the mineral nutrient contents in pods were found higher in soilless plants than in those grown in soil. In peas, root growth was better in plants grown on cocopeat than the plants grown under soil. However, when it comes to nodule formation, a good number of nodules was observed in the roots of plants grown on soil but no nodulation in the roots of the pea plants grown on cocopeat [30].

17.6 Cucumber

Arka Fermented Cocopeat recorded better growth and the highest yield of cucumber compared to soil [31]. Alifar *et al*. [32] also recorded higher biomass and cucumber fruit yield in cocopeat.

17.7 Tomato

Among the substrates studied, tomato plants raised on Arka Fermented Cocopeat registered maximum growth and yield (87.6 t/ha) compared to commercial cocopeat (76.7 t/ha) and soil (58.2 t/ha). The fruit quality was better when tomato plants were grown on Arka Fermented Cocopeat compared to commercial cocopeat and soil [33]. Plants grown in cocopeat substrate produced a higher fruit number (5.2%) and total yield (0.7%) than that of rockwool substrate. Fruit size and fruit quality characters showed no significant differences within growing substrates [34].

18. Standardized NPK levels for soilless vegetable production

Liquid nutrient formulations for growing zucchini, colour cabbage, chilli, coriander, cucumber, French bean, peas, and tomato on Arka Fermented Cocopeat under open and polyhouse soilless culture have also been developed. Best nutrient scheduling under open conditions was found to register maximum growth and yield in polyhouse conditions as well.

18.1 Zucchini

Nutrient scheduling of 168 ppm N-NO_3, 16 ppm P, and 189 ppm K recorded maximum fruit length (24.12 cm), fruit girth (44.4 mm), fruit weight (335.6 g), and yield (5.71 kg/plant and 71.39 t/ha) under open conditions. The above-mentioned nutrient scheduling recorded maximum growth and zucchini fruit yield in protected conditions also [25, 26].

18.2 Colour/red cabbage

Nutrient scheduling of 185 ppm N-NO_3, 41 ppm P, and 210 ppm K recorded maximum stem diameter (25.71 mm), a number of leaves (24.82), head diameter (36.79 cm), head length (14.64 cm), average head weight (972.25 g/plant), and yield (54.01 t/ha). The best nutrient scheduling under protected conditions is also found to register maximum growth and red cabbage yield in open conditions [27].

18.3 Chilli

In Chilli hybrid Arka Meghana, the highest number of fruits (248.2) and yield per plant (1.43 kg) was recorded with scheduling of 176 ppm N-NO_3, 29 ppm P, and 200 ppm K per plant and found to be on par with 194 ppm N-NO_3, 32 ppm P, and 228 ppm K (218.6 fruits and 1.30 kg yield per plant). However, the maximum fruit length (12.22 cm), fruit girth (10.98 mm), average fruit weight (5.96 g per fruit), and dry chilli yield (287 g per plant) was recorded with 194 ppm N-NO_3, 32 ppm P, and 228 ppm K nutrient scheduling [28].

18.4 Coriander

The production technology for soilless cultivation of coriander var. *Arka Isha* under open as well as in protected conditions has been standardized. The highest leaf yield of 4.79 t/ha was recorded with scheduling of 132 ppm N, 21 ppm P, and 150 ppm K under an open-field system of soilless cultivation. The performance of coriander under open field soilless culture was found to be better than polyhouse soilless culture [28].

18.5 Cucumber

Supplying of 166 ppm N-NO_3, 33 ppm P, and 207 ppm K recorded the maximum stem girth (18.43 mm), highest fresh (1690 g/plant), and dry plant biomass (540.8 g/

plant), highest average fruit weight (212.9 g) and yield (2.11 kg/plant and 32.51 t/ha) under open-field conditions [31].

18.6 French bean

Scheduling 141 ppm $N\text{-}NO_3$, 29 ppm P, and 179 ppm K recorded maximum plant height (47.11 cm), stem diameter (11.22 mm), number of branches (6.14), highest total fresh (205.8 g/plant), and dry biomass (35.89 g/plant), highest number of pods (42.74), pod length (15.04 cm), pod girth (7.31 mm), average pod weight (6.69 g), and yield (286.4 g/plant and 20.18 t/ha) [29].

18.7 Garden peas

Nutrient scheduling of 133 ppm $N\text{-}NO_3$, 27 ppm P, and 168 ppm K recorded maximum plant height (65.66 cm), stem diameter (6.51 mm), number of branches (3.14), highest plant biomass (24.04 g/plant), number of pods (15.14), pod length (7.07 cm), pod girth (9.67 mm), average pod yield (83.25 g/plant and 1.17 kg/bag) under open-field soilless cultivation. The best nutrient scheduling found in open-field conditions recorded better growth and yield under poly-house also [30].

18.8 Tomato

The highest number of fruits (80.14) and yield (93.9 t/ha) of tomato hybrid Arka Rakshak was recorded with the split application of 15:35:15 percent of the recommended NPK (180:120:180 kg NPK/ha), during establishment to early flowering, followed by 12.5:12.5:12.5 percent application during fruit development and 72.5:52.5:72.5 percent application during harvest. Nutrient scheduling significantly improved the TSS while other quality parameters were not significantly enhanced [33].

Tomato, colour cabbage, zucchini, and peas in soilless cultivation

19. Analysis of solution, tissue, and media

Knowledge of the nutritional status of all components (nutrient solution, substrate/media, and plant tissues) of a soilless cultivation system is very much required to judge the success of fertilizer schedules with respect to plant nutrients availability and the plant tissue nutrient content and it also helps to identify the reasons of any deficiency and toxicity symptoms that may appear in plants. The costs of the information with respect to the nutritional status of all components are a form of assurance towards success. The nutrient solution in a recirculated hydroponics system of cultivation may be utilized for a few days (short use) to a few weeks (extended use). To extend the life of nutrient solution to a few weeks in recirculated soilless culture/hydroponics system, it is always better to analyse the solution periodically for pH, EC, and individual nutrient concentration. Based on the nutrient analysis, periodic replenishment or adjustment in nutrient solutions can be made using nutrient stock solutions. By doing so, the longevity of nutrient solutions can be extended and the cost required for buying nutrient solutions or soluble salts can be reduced. Total salt content estimation on daily basis will also give the status of the nutrient content in the solution even though this cannot substitute for comprehensive analysis [2].

To avoid toxicity and deficiencies of nutrients in recirculated solutions due to continuous variation in nutrient status, it is necessary to do solution analysis for complete control over nutrient management in liquid soilless culture. The frequent requirement of solution analysis in water-based soilless culture gives a reason for switching over to solid substrate-based soilless culture. In solid substrate-based soilless culture systems, the evenly balanced nutrient solution is given to plants at the time of irrigation. In this way, the problem of nutrient solution management in solid substrate-based soilless cultivation systems can be minimized. Also, by accurately weighing the soluble salts at the time of nutrient solution preparation, it is possible to make a very properly working solution.

Like nutrient solution and substrate analysis, tissue analysis (leaf petioles or blades and whole leaves) is also warranted for successful nutrient management in plants. Tissue analysis during the crop growth period provides the current status of nutrient content in plants. Based on the nutrient content in plants, the fertilizer program may be adjusted or modified for better plant growth and productivity in soilless cultivation. Nutrient data obtained through tissue analysis may also help in interpreting nutrient deficiency or toxicity symptoms. Depending on plant parts sampled, location of sampling, and method used for analysis, the critical nutrient levels may vary. Critical nutrient concentrations for tomatoes, cucumbers, and different vegetables have been reported by various researchers [2, 35–37].

20. Water management in soilless vegetable cultivation

Substrate texture, porosity, and surface area to be wetted are vital considerations in making the right choice of irrigation in soilless vegetable cultivation [15]. While selecting an irrigation system for container or bag culture, one should keep in mind that the main purpose of irrigation is to apply nutrient solution homogeneously by making wet of entire growing media. A dry substrate or medium will make it very difficult for the plant root system to function properly [38]. Therefore, proper water management in soilless culture is very much important not only for meeting the water requirement of the plants but also for distributing the nutrients uniformly in the media. During summer, plants need extra water and hence the plants should preferably be irrigated twice a day. For soilless media, watering needs to be done only when the surface/subsurface of the media/substrate is dried and excess watering may be avoided.

21. Nutritional quality of vegetables grown in soilless culture

The results of most of the experiments conducted with different vegetable crops under open-field and polyhouse soilless culture indicated that the fruits of plants grown in soilless culture recorded better quality, particularly in mineral nutrient content compared to soil-grown plants [7, 33]. Most of the nutrient concentrations in zucchini fruits were found to be higher in soilless plants than in those grown in soil [7]. The fruit quality was better when tomato plants were grown on Arka Fermented Cocopeat compared to commercial cocopeat and soil.

Calcium content in tomato fruit samples was found to vary significantly among soilless media *viz,* cocopeat, rice hull, perlite, zeolite, and mica, and no significant variation was recorded with a phosphorus content of the fruits [39, 40]. Another study conducted by Borji *et al.* [41] revealed that there was no significant variation in Ca and Mg concentrations in tomato fruits obtained from the plants grown in different substrates.

The substrate combinations, *viz*., volcanic tuff, peat + volcanic tuff (1:1), volcanic tuff + spent mushroom compost (1:1), peat + volcanic tuff + spent mushroom compost (1:1:1), and soil, were compared for fruit ascorbic acid content in tomato and found no significant difference among growing media and soil [42]. Higher total soluble solid in tomato fruit produced from the tuff or sand-growing medium in two seasons was recorded as compared to fruit growing in soil [43]. A comparison, between rockwool (R), perlites plus carbonized rice hull (PCRH), cypress bark (CB), and coconut coir (CD) was carried out in a greenhouse using a small type of tomato (*Lycopersicon esculentum* Mill. cv. T-148) in a summer experiment. The high total soluble solids content (°Brix) was represented by coconut coir (CD) treatment [44]. Most of the macro and micronutrient concentrations in cucumber fruits were found to be higher in soilless plants than in those grown in soil. The highest K, Ca, and Cu were recorded in cucumber fruits harvested from the plants grown on soil. Most of the macro and micronutrient concentrations in French bean pods were found to be higher in soilless plants than in those grown in soil. However, the highest K was recorded in the pods harvested from the plants grown on soil compared to Arka Fermented Cocopeat.

22. Nematode infection and nodulation under soilless cultivation

In polyhouse French bean cultivation, nematode infection was found to be almost nil in plants grown on cocopeat but nearly half of the plants grown in soil were affected with a nematode [29]. In peas, root growth was better in plants grown on cocopeat than the plants grown under soil. However, when it comes to nodule formation, a good number of nodules was observed in the roots of plants grown on soil but no nodulation in the roots of the pea plants grown on cocopeat [30].

23. Net profit from soilless vegetable cultivation

The results of most of the experiments conducted at ICAR-IIHR, Bengaluru with different vegetable crops in grow bags under open-field and polyhouse soilless culture indicated that the plants grown in soilless culture recorded higher yield and better quality, particularly in mineral nutrient content compared to soil-grown plants. The yield of different vegetables grown under soilless culture in an area of 100 m^2 is as follows; 1260 kg for tomato, 803.6 kg for zucchini, 204 kg for colour cabbage, 300 kg for chilli, 441 kg for cucumber 280 kg for French bean, and 81.9 kg for garden peas. Net profit from the vegetables grown in an area of 100 m^2 varied from Rs 7140 for cucumber to Rs 35,960 for zucchini and the net profits of the rest of the crops found to fit in between. This technology would be highly suited for urban and peri-urban vegetable cultivation for meeting the food security in cities. The production technology developed at ICAR-Indian Institute of Horticultural Research for soilless cultivation of most commonly consumed vegetables in India has generated a lot of interest among the soilless growers for the cultivation of vegetables on AFC. This particular technology is being popularized through various training programmes, exhibitions, magazines, and media. Many growers have already started adopting this particular technology in the cultivation of vegetables using AFC as a substrate.

24. Advantages and constraints of soilless cultivation

A substantial quantum of research work carried out in recent past stating the advantages and disadvantages of soilless cultivation of vegetables.

24.1 Advantages

Compared to a conventional soil-based cultivation system, soilless cultivation provides several advantages than disadvantages. Soilless cultivation provides ideal conditions for the growth of plants which in turn helps in getting a higher yield. With little effort, time and cost, it is possible to do very relaxed and clean vegetable cultivation under soilless culture. The majority of soil-born pests and diseases can be controlled just by shifting over to soilless cultivation from the traditional way of farming. Degraded and poor fertile soils can be easily brought into soilless cultivation. It affords an unsoiled working environment and thus labour engagement is easy. List of other advantages of soilless culture is control of plant nutrition, ability to control pH and EC, water economy and control, reduction of labour requirement, sterilization practices, control of root environment, multiple crops per year and unsuitable soil can be used, etc.

24.2 Limitations

In spite of several merits offered by soilless culture, it has few demerits as well. Technical know-how and high initial cost are the two important things required for scaling up of soilless culture at the commercial level. The requirement of investment and technical knowledge will go up further when combining soilless culture with protected cultivation. Experts with precision management skills are needed for nutrient solution preparation, pH and EC maintenance, identification and correction of mineral nutrient deficiency, aeration; upkeeping all the weather parameters in support of ideal plant growth in protected structures, etc. Above all, much attention is important for plant health management. The requirement of energy inputs is very high to run the soilless culture system, particularly in hydroponics. Because of higher initial cost, technical knowledge on crop agronomy and physiology limits the soilless culture to high-value crops cultivation. Growing low-value crops in hydroponics may not be so economical.

25. Conclusion

In urban and peri-urban agriculture, no doubt that the soilless culture is rapidly gaining impetus and acceptance among growers. In advanced countries, the system of soilless cultivation is so popular and well-received mainly for commercial cultivation of high-value vegetable crops, medicinal and ornamental crops but now it is spreading very rapidly in rest of the world. With this speed, the soilless culture is certainly going to dominate in future food production. Growers are presently turning towards alternate technologies, such as soilless culture due to the decline in the availability of arable lands and the problem of soil-borne diseases in soil-based cultivation. Due to better water use efficiency in soilless culture, this particular system of cultivation can also be taken to places where water availability is limited. Presently the hydroponics unit setting up cost is too high because of limited adoption but by acceptance and adoption of more and more growers, the cost of the unit can be brought to affordable levels. Further, this technology is not getting popularity as expected in some of the developing and underdeveloped countries due to various reasons, such as high initial investment and the requirement of skilled manpower. Standardized soilless production technology by the public and private research institutions is very important to popularize and create mass awareness among urban and peri-urban growers. In this direction, ICAR-IIHR is not stopping after standardizing the soilless production technology for vegetables but also putting more and more effort into the spread of this particular technology at the national level. ICAR-IIHR soilless culture production technology has already been popularized through various training programmes, exhibitions, magazines, and media. Many growers have already started adopting IIHR soilless culture technology in the cultivation of vegetables, flowers, and medicinal crops.

Author details

Duraisamy Kalaivanan*, Govindan Selvakumar
and Arockiasamy Carolin Rathinakumari
ICAR-Indian Institute of Horticultural Research, Bengaluru, India

*Address all correspondence to: kalainith2003@gmail.com

References

[1] Kalaivanan D, Selvakumar G. Soilless cultivation of vegetables and ornamental crops. Kisan World. 2016;**43**(7):23-26

[2] Johnson H Jr. Hydroponics: A Guide to Soilless Culture. Leaflet 2947. Berkeley: Division of Agriculture and Natural Resources, Univ. of California; 1980

[3] Kalaivanan D, Selvakumar G, Ganeshamurthy AN, Shankara Hebbar S. Soilless culture of vegetables. In: Book Abstracts of SciCon Series 3rd International Conference on In-Sync with Next-Generation Biosciences (INGB)-2019; 06-08 November 2019; Hotel Cochin, Kerala, India: Scire Science; 2019. p. 9

[4] Kalaivanan D, Selvakumar G, Ganeshamurthy AN, Shankara Hebbar S. Soilless culture: An avenue for urban and peri-urban vegetable production. In: Book Abstracts of National Conference on Recent Advances in Smart Gardening and Landscaping; 23rd and 24th November 2019; Mumbai: Mumbai Kalina Campus Auditorium, Mumbai University; 2019

[5] Kalaivanan D, Selvakumar G, Ganeshamurthy AN, Shankara Hebbar S. Soilless culture: An avenue for urban and peri-urban vegetable production. In: Published in Compendium of State Level Seminar on Roof top Gardening and Urban Horticulture; 24th and 25th November 2019; IIHR-CHES, Bhubaneswar: Rotary Club of Bhubaneswar Heritage; 2019

[6] Singh S, Singh BS. Hydroponics—A technique for cultivation of vegetables and medicinal plants. In: Proceedings of 4th Global Conference on Horticulture for Food, Nutrition and Livelihood Options; 28-31 May 2012; Bhubaneshwar, Odisha, India: Westville Publishing House; 2012. p. 220

[7] Kalaivanan D, Selvakumar G, Ganeshamurthy AN. Vertical farming—an avenue for urban horticultural production. Kisan World. 2017;**44**(5):45-46

[8] Gericke WF. The Complete Guide to Soilless Gardening. New York: Prentice Hall; 1940

[9] Boodley JW, Sheldrakejr R. Cornell peatlite mixes for commercial plant growing. In: Informational Bulletin 43. Ithaca, New York: New York State College of Agriculture and Life Sciences; 1977

[10] Kalaivanan D, Selvakumar G, Nair AK, Sridhar V, Sriram S. Terrace Gardening for fresh, green, clean and local vegetable production. Agro India. 2021;**6**:33-36

[11] Csaba I. Growing medium in hydroculture. Plasticulture. 1995;**108**(4):45-47

[12] Olympios CM. Soilless media under protected cultivation rockwool, peat, perlite and other substrates. Acta Horticulturae. 1995;**401**:443-451

[13] Olympios CM. Overview of soilless culture: Advantages, constraints and perspectives for its use in Mediterranean Countries. Cahiers Options Méditerranéennes. 1999;**31**:307-324

[14] Berjón MA, Murray PN, Benedito CC. Substrates for soilless culture and fertigation. In: Cadahía C, editor. Fertigation: Vegetable Crops, Fruit and Ornamental. Madrid, Spain: Mundi-Prensa; 2005. pp. 299-354. In Spanish

[15] Chen Lopez JC, Waller P, Giacomelli G, Tuller M. Physical characterization of greenhouse substrates for automated irrigation management. Acta Horticulturae. 2008;**797**:333-338

[16] Hassall and Associates. Hydroponics as an Agricultural Production System: A report for the Rural Industries Research and Development Corporation. Publié en Novembre 2001 (Publication de RIRDC No 01/141) (Projet RIRDC No HAS-9A). 2001

[17] Jiang WJ, Yu HJ. Twenty years development of soilless culture in Mainland China. Acta Horticulturae. 2007;**759**:181-186

[18] Donnan R. Hydroponics around the world. Practical Hydroponics & Greenhouses. 1998;**41**:18-25

[19] Bradley P, Marulanda C. Simplified hydroponics to reduce global hunger. Acta Horticulturae (ISHS). 2001;**554**:289-296. DOI: 10.17660/ActaHortic.2001.554.31

[20] Cooper A. The ABC of NFT: The World's First Method of Crop Production Without a Solid Rooting Medium. London, UK: Grower Books; 1979. pp. 179-181

[21] Steiner AA. The universal nutrient solution. In: 6th International Congress on Soilless Culture. Wageningen, The Netherlands: ISOSC; 29 April–5 May 1984. pp. 633-650

[22] Hewitt EJ. Sand and water culture methods used in the study of plant nutrition. In: Hewitt EJ, editor. Farnham Royal. England: Commonwealth Agricultural Bureaux; 1966. p. 547

[23] Hoagland DR, Arnon DI. The Water-Culture Method for Growing Plants Without Soil. University of California, College of Agriculture, Agricultural Experiment Station; 1938

[24] Carrijo OA, Vidal MC, Reis NVB, Souza RB, Makishima N. Tomato crop production under different substrates and greenhouse models. Horticultura Brasileira. 2004;**22**:5-9. In Portuguese, with abstract in English

[25] Kalaivanan D, Selvakumar G, Ganeshamurthy AN, Shankara Hebbar S. Soilless cultivation of zucchini in open and protected conditions. Kisan World. 2017;**44**(8):53-54

[26] Kalaivanan D, Selvakumar G, Shankara Hebbar S. Effects of varying N, P and K concentrations on growth, biomass, yield and nutritional quality of zucchini squash grown under open and polyhouse soilless culture. Indian Journal of Horticulture. 2020;**77**(3):496-502. DOI: 10.5958/0974-0112.2020.00071.7

[27] Kalaivanan D, Selvakumar G, Ganeshamurthy AN, Shankara Hebbar S. Soilless cultivation of red cabbage under open-field and protected conditions. Kisan World. 2017;**44**(10):52-53

[28] Kalaivanan D, Selvakumar G, Ganeshamurthy AN, Shankara Hebbar S. An open-field soilless culture of chilli and coriander on Arka Fermented Cocopeat. Kisan World. 2018;**45**(12):45-47

[29] Kalaivanan D, Selvakumar G, Ganeshamurthy AN, Shankara Hebbar S. Open and polyhouse soilless culture of French bean var. Arka Sharath on Arka Fermented Cocopeat. Kisan World. 2018;**46**(3):49-50

[30] Kalaivanan D, Selvakumar G, Ganeshamurthy AN, Shankara Hebbar S. Garden Peas: Arka Apoorva Cultivation Under Open-Field and Polyhouse Soilless Culture. Kisan World. 2019;**46**(4):39-42

[31] Kalaivanan D, Selvakumar G, Ganeshamurthy AN, Shankara Hebbar S. An outdoor soilless culture of cucumber on Arka Fermented Cocopeat. Kisan World. 2019;**46**(4):49-50

[32] Alifar N, Mohammadi Ghehsareh A, Honarjoo N. The Effect of growth media on cucumber yield and its uptake of some nutrient elements in soilless culture. Journal of science and technology Greenhouse Culture. 2010;**1**:19-25

[33] Kalaivanan D, Selvakumar G, Ganeshamurthy AN, Shankara Hebbar S. Polyhouse production of tomato on Arka Fermented Cocopeat. Kisan World. 2016;**43**(11):12-14

[34] Binod PL, Prakash Babu A, Cheol Soo Y, Won Hee K. Yield and fruit quality of tomato (Lycopersicon esculentum Mill.) cultivars established at different planting bed size and growing substrates. Horticulture, Environment, and Biotechnology. 2012;**53**(2):102-107

[35] Ward GM. Leaf analysis of vegetable crops. In: Agdex 290/5 32. Ontario, Canada: Ontario Ministry of Agriculture; 1973

[36] Witmer SH, Honma S. Greenhouse Tomatoes, Lettuce, and Cucumbers. East Lansing: Michigan State Univ. Press; 1979

[37] Lorenz OA, Tyler KB. Plant tissue analysis of vegetable crops. In: Soil and Plant Testing in California. Bulletin 1879. Berkeley: Division of Agriculture and Natural Resources, Univ. of California; 1983

[38] Ismail SM, Ozawa K, Khondaker NA. Influence of single and multiple water application timings on yield and water use efficiency in tomato (variety First power). Agricultural Water Management. 2008;**95**:116-122

[39] Saberi Z. Use of zeolite, vermiculate and some inert materials as media for hydroponic tomato production [Master's Thesis Soil Science]. Isfahan, Iran: Isfahan University of Technology; 2009

[40] Djedidi M, Gerasopoulos D, Maloupa E. The effect of different substrates on the quality of *F. Carmello* tomatoes grown under protection in a hydroponics system. Cahiers Option Mediterranean's. New Delhi, India: Westville Publishing House; 2001;**31**:379-383

[41] Borji H, Mohammadi-Ghehsareh A, Jafarpour M. Effects of the substrate on tomato in soilless culture. Research Journal of Agricultural Biologic Science. 2010;**6**(6):923-927

[42] Celikel G. Effect of different substrates on yield and quality of tomato. Acta Horticulturae. 1999;**491**:353-356

[43] Fandi M, Al-Muhtaseb JA, Hussein MA. Yield and fruit quality of tomato as affected by the substrate in an open soilless culture. Jordan Journal of Agricultural Sciences. 2008;**4**:65-72

[44] Inden HA, Torres. Comparison of four substrates on the growth and quality of tomatoes. Acta Horticulturae (ISHS). 2004;**644**:205-210

Chapter 7

Intensive Production of *Solanum lycopersicum* in Soil and Greenhouse

Víctor García-Gaytan and Fanny Hernández-Mendoza

Abstract

S. lycopersicom plants are one of the most cultivated vegetables, and their fruits are consumed worldwide. The cultivation system can be carried out in soil and hydroponics. Its cultivation in soil must be planned properly. Within this, the selection of the variety of interest, the germination, and care of these stand out. The preparation of the land, plot, or farm, the physical–chemical analysis of the soil for the application of amendments (minerals, compost, and vermicompost). The mechanism of the hydraulic irrigation system for the application of water and nutrients. The mulches system, care, and management of plants during their growth. The application of biostimulants and the frequency of fertigation. In this chapter, we will address interesting topics for the management of high tomato production in greenhouse soil.

Keywords: vegetal nutrition, variety, physiochemical, minerals, fertigation

1. Introduction

For the intensive production of tomato, it must include the selection of the variety of interest, the germination of seeds in a certified nursery, the care of the seedlings in which it must include irrigation, temperature measurements, pest and disease control. And above all, take care of the height of the seedling, as it is a determining factor for good development in the field. Very tall seedlings present stress during planting in soil. Losses during the transplant process can reach 10–15%.

Before germinating the seeds of interest in the nursery, it is necessary to consider the preparation of the soil. During soil preparation, it is essential to carry out a physical–chemical analysis of the soil. This analysis should include: availability of primary and secondary macronutrients (ppm), organic matter (OM), EC, pH, bulk density (BD), and cation exchange capacity (CEC). Also, it will be necessary to carry out an analysis of the water quality. This analysis must include: pH, EC, the presence of bicarbonates (meq L^{-1}), and take actions for its displacement with acids (HNO_3, H_2SO_4, and H_3PO_4), to lower the pH, provide nitrogen, phosphorus, and remove salt from the irrigation system. With the soil analysis, amendments will be applied (t ha^{-1}), application of OM (compost and vermicompost), agricultural plaster, and agricultural lime.

In addition, it is necessary to know and monitor the average temperature in each of the stages of the plant (germination – growth – flowering – fruiting). Temperature should include daytime and nighttime. This variable is extremely important because the success of production depends on it. In this chapter, we will address the importance of the available technologies for the intensive production of *Solanum lycopersicum* in soil.

2. Soil preparation

To achieve success in *S. lycopersicum* production megafactories, planning well in advance is essential. Among them, the preparation of the land, plot, or farm for cultivation stands out. Tracking and fallow by means of adequate machinery are determining factors for anchoring and root distribution. Once the land has been prepared, it will be necessary to carry out soil sampling. Soil subsamples can be taken in Zig-Zag or with another methodology according to the ground conditions. The variables of the physical–chemical analysis mentioned above will help to manage a good development of the crop in each growth stage [1]. Once the laboratory sends the results to the users, those responsible for the farm will take action to manage the soil. These actions may include applications of amendments, vermicompost, or mineral fertilizers (personal communication).

3. Mineral fertilizer

Mineral fertilizers help plant roots to develop in saline conditions by increasing soil porosity, nutrient content, and organic matter. The integration of mineral fertilizers, plus the application of nutrient solutions, will have a positive impact on root biomass (**Figure 1**).

The organo-mineral fertilizer may acts as a reservoir for nutrients, ensuring slow release to the substrate solution or directly to plant roots, it is a relatively abundant mineral resource [2]. The vermicompost improves microbial diversity and abundance, plant growth, and soil fertility [3]. **Table 1** shows the effect of organ-mineral fertilizer and vermicompost on different crops.

Figure 1.
Root biomass. A) Nutrient solution. B) Granulated mineral fertilizer + nutrient solution. C) Powdered mineral fertilizer + nutrient solution. (Courtesy Dr. Víctor García-Gaytán).

Crops	Effect on the crop	Reference
Cucumis sativus L.	The organo-mineral fertilizer decrease soil salinity (EC), soil pH, and increase field capacity, total porosity, soil nutrients content and soil organic matter content.	[4]
Solanum melongena L.	The organo-mineral fertilizer plants showed increased growth, proline, chlorophyll, nutrient contents, total porosity, and field capacity.	[5]
Solanum tuberosum L.	The application of 1.5 t ha^{-1} of organo-mineral fertilizer at planting, the performance of the tuber was improved.	[6]
Phaseolus vulgaris L.	A partial alternative to chemical fertilizers demonstrated the validity and possibility of sustainable agronomic performance.	[7]
Sorghum bicolor L.	Organic compost and soil mulching improved seed and forage yield production	[8]
S. tuberosum, Spinacia olearecea, Brassica campestris	Vermicompost significant improvement in the soil quality. In spinach, the requirement is 4 t ha^{-1}, while in potato and turnip, it was higher 6 t ha^{-1}.	[9]

Table 1.
Benefits of the application of organo-mineral fertilizer and vermicompost in different crops.

4. Irrigation lines and mulches

Using suitable machinery, it will be necessary to make the bedding (furrows or edges). The main irrigation lines (pipes: hydraulic PVC) are the ones that will carry the greatest flow of water to the irrigation sections. In each irrigation section, the secondary irrigation lines (irrigation belt) are installed, these lines are the ones that will be on the edges [1]. Once the entire irrigation system is installed, it is time to start installing the plastic mulches on the furrows (personal communication). The emitters

Figure 2.
Bedding system: Irrigation strap, plastic, and tomato plants grown in soil with high sodium concentration. (Courtesy Dr. Víctor García-Gaytán).

in the strips are responsible for efficiently distributing the water in each irrigation section. The amount of water applied to each plant will be calculated based on the irrigation flow rate, soil type, demand, and growth stage of the tomato (personal communication). For industrial production dimensions, it will be necessary to install the best brand of irrigation belt that the market offers, which provides durability and resistance to the pressure exerted by the irrigation flow (personal communication) (**Figure 2**).

5. Nutrient solution in fertigation

The application of irrigation and nutrition will be monitored by qualified personnel (personal communication). The monitoring will be carried out from the irrigation room, where the semiautomated/automated irrigation head will be installed. The farm manager must have computer skills to program the nutrient application events/times in the irrigation sections (personal communication). In nutrient injection systems, it is essential to have a container for the application of special solutions (biostimulants: seaweed extract, protein hydrolyzates, humic acids, phosphites, bacteria, and fungi). Special solutions are to correct nutritional deficiencies, stress, or control pests and diseases quickly and effectively. The dose, handling, mixing, and application must be supervised/monitored by experts in plant nutrition (personal communication).

High yield and product quality of crops are only possible if nutrition is optimized, this includes nutrient solution composition, water supply, nutrient solution temperature, dissolved oxygen concentration, electrical conductivity, and pH of the nutrient solution [10]. Oxygen content in nutrient solutions has an effect on the yield of sweet pepper and melon about 20% and 15% [11]. EC is related to the amount of ions available to plants in the root zone [12]. pH and EC monitoring helps to eliminate the problem associated with fertilization, showing the nutrient availability to the crops [13]. There must be a mutual relationship between the anions and cations of the nutrient solution [14]. The nutrient interactions are generally measured in terms of growth response and change in the concentration of nutrients [15]. Proper knowledge of possible synergistic and antagonistic interactions between nutrients [16] is required. Differences in nutrients concentrations between desired levels in the solutions and those found in the tap water represent the amounts of nutrients per volume nutrient solution (meq L^{-1}), which should be added through fertilizers [17]. The efficient acquisition of water and mineral elements y plants roots is a prerequisite for sustainable intensification of crop production [18]. Root growth is important for effective exploration of soil and interception of nutrients [19]. Interaction with microorganisms is also significant [20]. When the roots absorb excessive cation compared with anions, the roots offset this by excreting protons (H^+), which generally leads to rhizosphere acidification. When they absorb more anions than cations, the roots excrete hydroxyl (OH^-). Hydroxyl reacts with carbon dioxide to form bicarbonate (HCO_3^-), which leads to rhizosphere alkalization [21]. The calcium levels $Ca(NO_3)_2$ as a soluble fertilizer and nitrogen should be considered in the irrigation systems; in addition, it should be considered that the Ca^{2+} in the soil solution competes with K^+, Mg^{2+}, and N [22]. To achieve high-quality yield, a balance of nutrients in the rhizosphere must be maintained for each stage of growth [23]. The manipulation of the climate of the greenhouse (RH, temperature, level of environmental CO_2) can compensate the negative effects of high salinity on the yield and quality of the fruit [23].

5.1 Monitoring of soil and fertigation frequency

The monitoring of soil moisture, the frequency of fertigation, the proper functioning of the emitters should be checked continuously, to prevent the accumulation of salts on the surface of the soil and prevent nutritional imbalances. The surface layer of the soil tends to accumulate high concentrations of salts after evaporation or transpiration, when the irrigation water contains soluble salts [24]. Monitoring the saturated extract medium provides an assessment of both the applied salinity load and the concentration of fertilizer salts [24]. Transpiration during the day may cause significant differences between the water content in the rhizosphere and that in the bulk soil [25]. Selective electrode analysis of NO_3-N concentration of soil solution sample obtained by suction lysimetry, it is a good monitoring method [26].

Soil testing to monitor nutrient availability in the root zone is a valuable alternative [27]. The distribution of NO_3-N in the soil profile is very greatly influenced by fertigation frequency in sandy-loam soil [28]. Soil-moisture sensor-based irrigation system in bell pepper significantly reduced the applied irrigation resulting in 7–62% less irrigation water applied compared with fixed time irrigation treatment without compromising marketable yield [29]. Smart irrigation, where a well-scheduled and well-dosed irrigation regime is essential, as by applying the right amount of water at the right time, one avoids plant demand being either exceeded or not met [30]. Despite the potential of fertigation for effective water and N saving, in practice, there is a need for improvement in the management of this technology to make it effective [31]. Increasing the irrigation frequency significantly enhanced the transpiration flux so that the transpiration flux of plants under low irrigation-P level at 10 daily irrigation events was similar to that of plants of lettuce under high solution-P [32].

5.2 Bioprotection of roots

Optimization is needed in the nutrition of crops, which involves the use of biostimulants to counter oxidative stress and the management of strain bioformulations (bacteria and fungi) that protect and stimulate root for the acquisition of nutrients [1]. Biostimulant action is diverse, but it can include N metabolism activation, P release on soil, stimulation of soil microbial activity, and root stimulation [33]. These include seaweed extract [34, 35], protein hydrolyzates [36, 37], humic acids [38, 39], phosphites [40, 41]. Bacteria and fungi also stimulate and protect the roots [1], these include beneficial microorganisms, which can be free-living, rhizospheric, or endosymbiotic [42]. According to [43], specialized plant membrane transporters can be used to enhance yields of staple crops, increase nutrient content, and increase resistance to key stresses, including salinity, pathogens, and aluminum toxicity.

6. Conclusion

In this chapter, tools for the intensive cultivation of *S. lycopersicom* are obtained. Its cultivation in soil must be planned properly. Within this, the selection of the variety of interest, the germination, and care of these stand out. The preparation of the land, the physical–chemical analysis of the soil for the application of amendments (minerals, compost, and vermicompost). The mechanism of the irrigation system for the application of water and nutrients. The mulches system, care, and management of plants during their growth. The application of biostimulants and the frequency of fertigation.

Conflict of interest

The authors declare no conflict of interest in this manuscript.

Author details

Víctor García-Gaytan[1*] and Fanny Hernández-Mendoza[2]

1 Plant Nutrition and Horticulture, Mexico, Mexico

2 Independent Researcher Comprehensive Improvement of Plants through Genetics and Plants Nutrition, Mexico, Mexico

*Address all correspondence to: dr.garvictan@yahoo.com; garvictan@gmail.com

References

[1] García-Gaytán V, Hernández-Mendoza F, Coria-Téllez AV, García-Morales S, Sánchez-Rodríguez E, et al. Fertigation: Nutrition, stimulation and bioprotection of the root in high performance. Plants. 2018;7:88

[2] Rady MM. A novel organo-mineral fertilizer can mitigate salinity stress effects for tomato production on reclaimed saline soil. South African Journal of Botany. 2012;**81**:8-14

[3] Maji D, Misra P, Singh S, Kalra A. Humic acid rich vermicompost promotes plant growth by improving microbial community structure of soil as well as root nodulation and mycorrhizal colonization in the roots of *Pisum sativum*. Applied Soil Ecology. 2017;**110**:97-108

[4] El-Mageed TAA, Semida WM. Organo mineral fertilizer can mitigate water stress for cucumber production (*Cucumis sativus* L.). Agricultural Water Management. 2015;**159**:1-10

[5] Semida WM, Abd El-Mageed TA, Howladar SM. A novel organo-mineral fertilizer can alleviate negative effects of salinity stress for eggplant production on reclaimed saline calcareous soil. International Symposium on Growing Media and Soilless Cultivation. 2013;**1034**:493-499

[6] Zebarth BJ, Chabot R, Coulombe J, Simard RR, Douheret J, et al. Pelletized organo-mineral fertilizer product as a nitrogen source for potato production. Canadian Journal of Soil Science. 2005;**85**:387-395

[7] Rady MM, Semida WM, Hemida KA, Abdelhamid MT. The effect of compost on growth and yield of Phaseolus vulgaris plants grown under saline soil. International Journal of Recycling Organic Waste and Agriculture. 2016;**5**:311-321

[8] El-Mageed TAA, El-Samnoudi IM, Ibrahim AE-AM, El Tawwab ARA. Compost and mulching modulates morphological, physiological responses and water use efficiency in sorghum (*Bicolor* L. Moench) under low moisture regime. Agricultural Water Management. 2018;**208**:431-439

[9] Ansari AA. Effect of vermicompost on the productivity of potato (*Solanum tuberosum*) spinach (*Spinacia oleracea*) and turnip (*Brassica campestris*). World Journal of Agricultural Science. 2008;**2008**(4):333-336

[10] Gorbe E, Calatayud A. Optimization of nutrition in soilless systems: A review. Advances in Botanical Research. Elsevier. 2010:193-245

[11] Urrestarazu M, Mazuela PC. Effect of slow-release oxygen supply by fertigation on horticultural crops under soilless culture. Scientia Horticulturae. 2005;**106**:484-490

[12] Ding X, Jiang Y, Zhao H, Guo D, He L, et al. Electrical conductivity of nutrient solution influenced photosynthesis, quality, and antioxidant enzyme activity of pakchoi (*Brassica campestris* L. ssp. *Chinensis*) in a hydroponic system. PLoS One. 2018;**13**:e0202090

[13] Ludwig F, Fernandes DM, Mota PR, Bôas RLV. Electrical conductivity and pH of the substrate solution in gerbera cultivars under fertigation. Horticultura Brasileira. 2013;**2013**(31):356-360

[14] Steiner AA. A universal method for preparing nutrient solutions of a certain desired composition. Plant and Soil. 1961;**15**:134-154

[15] Fageria VD. Nutrient interactions in crop plants. Journal of Plant Nutrition. 2001;**24**:1269-1290

[16] Rietra RP, Heinen M, Dimkpa CO, Bindraban PS. Effects of nutrient antagonism and synergism on yield and fertilizer use efficiency. Communications in Soil Science and Plant Analysis. 2017;**48**:1895-1920

[17] Savvas D, Adamidis K. Automated management of nutrient solutions based on target electrical conductivity, pH, and nutrient concentration ratios. Journal of Plant Nutrition. 1999;**22**:1415-1432

[18] White PJ, George TS, Gregory PJ, Bengough AG, Hallett PD, et al. Matching roots to their environment. Annals of Botany. 2013;**112**:207-222

[19] Lynch J. Root architecture and plant productivity. Plant Physiology. 1995;**109**:7

[20] Richardson AE, Barea J-M, McNeill AM, Prigent-Combaret C. Acquisition of phosphorus and nitrogen in the rhizosphere and plant growth promotion by microorganisms. Plant and Soil. 2009;**321**:305-339

[21] Mmolawa K, Or D. Root zone solute dynamics under drip irrigation: A review. Plant and Soil. 2000;**2000**(222):163-190

[22] Mayfield JL, Kelley WT. Blossom-End Rot and Calcium Nutrition of Pepper and Tomato. UGA Cooperative Extension; 2009

[23] Dorai M, Papadopoulos A, Gosselin A. Influence of Electric Conductivity Management on Greenhouse Tomato Yield and Fruit Quality. EDP Sciences; 2001

[24] Jacobs DF, Timmer VR. Fertilizer-induced changes in rhizosphere electrical conductivity: Relation to forest tree seedling root system growth and function. New Forest. 2005;**30**:147-166

[25] Silber A, Xu G, Levkovitch I, Soriano S, Bilu A, et al. High fertigation frequency: The effects on uptake of nutrients, water and plant growth. Plant and Soil. 2003;**253**:467-477

[26] Hartz TK, Smith RF, LeStrange M, Schulbach KF. On-farm monitoring of soil and crop nitrogen status by nitrate-selective electrode. Communications in Soil Science and Plant Analysis. 1993;**24**:2607-2615

[27] Incrocci L, Massa D, Pardossi A. New trends in the fertigation management of irrigated vegetable crops. Horticulturae. 2017;**3**:37

[28] Rajput T, Patel N. Water and nitrate movement in drip-irrigated onion under fertigation and irrigation treatments. Agricultural Water Management. 2006;**79**:293-311

[29] Zotarelli L, Dukes MD, Scholberg JMS, Femminella K, Munoz-Carpena R. Irrigation scheduling for green bell peppers using capacitance soil moisture sensors. Journal of Irrigation and Drainage Engineering. 2010;**137**:73-81

[30] Gu Z, Qi Z, Ma L, Gui D, Xu J, et al. Development of an irrigation scheduling software based on model predicted crop water stress. Computers and Electronics in Agriculture. 2017;**143**:208-221

[31] Quemada M, Gabriel JL. Approaches for increasing nitrogen and water use

efficiency simultaneously. Global Food Security. 2016;**9**:29-35

[32] Xu G, Levkovitch I, Soriano S, Wallach R, Silber A. Integrated effect of irrigation frequency and phosphorus level on lettuce: P uptake, root growth and yield. Plant and Soil. 2004;**263**:297-309

[33] Yakhin OI, Lubyanov AA, Yakhin IA, Brown PH. Biostimulants in plant science: A global perspective. Frontiers in Plant Science. 2017;**2017**(7):2049

[34] Hamza B, Suggars A. Biostimulants: Myths and realities. TurfGrass Trends. 2001;**8**:6-10

[35] Stirk W, Arthur G, Lourens A, Novak O, Strnad M, et al. Changes in cytokinin and auxin concentrations in seaweed concentrates when stored at an elevated temperature. Journal of Applied Phycology. 2004;**2004**(16):31

[36] Colla G, Hoagland L, Ruzzi M, Cardarelli M, Bonini P, et al. Biostimulant action of protein hydrolysates: Unraveling their effects on plant physiology and microbiome. Frontiers in Plant Science. 2017;**8**

[37] Rouphael Y, Colla G, Giordano M, El-Nakhel C, Kyriacou MC, et al. Foliar applications of a legume-derived protein hydrolysate elicit dose-dependent increases of growth, leaf mineral composition, yield and fruit quality in two greenhouse tomato cultivars. Scientia Horticulturae. 2017;**226**:353-360

[38] Nikbakht A, Kafi M, Babalar M, Xia YP, Luo A, et al. Effect of humic acid on plant growth, nutrient uptake, and postharvest life of gerbera. Journal of Plant Nutrition. 2008;**31**:2155-2167

[39] Nardi S, Ertani A, Francioso O. Soil–root cross-talking: The role of humic substances. Journal of Plant Nutrition and Soil Science. 2017;**180**:5-13

[40] Dalio RJ, Fleischmann F, Humez M, Osswald W. Phosphite protects *Fagus sylvatica* seedlings towards *Phytophthora plurivora* via local toxicity, priming and facilitation of pathogen recognition. PLoS One. 2014;**9**:e87860

[41] Achary VMM, Ram B, Manna M, Datta D, Bhatt A, et al. Phosphite: A novel P-fertilizer for weed management and pathogen control. Plant Biotechnology Journal. 2017

[42] du Jardin P. Plant biostimulants: Definition, concept, main categories and regulation. Scientia Horticulturae. 2015;**196**:3-14

[43] Schroeder JI, Delhaize E, Frommer WB, Guerinot ML, Harrison MJ, et al. Using membrane transporters to improve crops for sustainable food production. Nature. 2013;**497**:60